Sarah's Search
A Silk Odyssey

Ian Braybrook

Marilyn Bennet

By the same Author

Gweneth Wisewould – Outpost Doctor
Six Ha'pennies
The Changing Times
Bush Wireless

Sarah's Search – A Silk Odyssey
First Published in 2017 by
Marilyn Bennet Publishing
Maine Media, PO Box 677, Castlemaine Vic 3450

Copyright © Ian Braybrook & Marilyn Bennet

All rights reserved. Without limiting the rights under copyright reserved above, no part of this publication may be reproduced, stored in or introduced into a retrieval system, or transmitted, in any form or by any means (electronic, mechanical, photocopying, recording or otherwise), without the prior written permission of both the copyright owner and Marilyn Bennet Publishing.

National Library of Australia
Cataloguing-in-Publication entry
Sarah's Search – A Silk Odyssey
Braybrook, Ian & Bennet, Marilyn, authors

ISBN 978-0-9944370-2-0

1.Title 2. Silk 3. Sericulture 4. Pioneer Women 5. Australian & Local History 6. Corowa 7. Harcourt 8. Mount Alexander 9. Horticulture 10. Agriculture 11. Mulberry

Text design and layout by Level Heading – levelheading.com

Edited by Bernard Schultz

Photographs courtesy of the authors and of the various sources acknowledged.

On notification, any inadvertent copyright omissions will be corrected in a future edition.

Supported by the Mount Alexander Shire Council Community Grants Program

This book is dedicated to
the late Hedley and Sybil James
who preserved the history of Harcourt and District

Contents

Foreword. v
About the Authors. vii
Acknowledgements viii
Prologue x
Introduction xiii
Terms used in Sericulture xviii
Chapter One 21
Chapter Two 25
Chapter Three 35
Chapter Four. 45
Chapter Five 49
Chapter Six. 53
Chapter Seven 58
Chapter Eight 62
Chapter Nine. 76
Chapter Ten 86
Chapter Eleven 92
Chapter Twelve 97
Chapter Thirteen 115
Chapter Fourteen 119
Chapter Fifteen 125
Chapter Sixteen 129
Chapter Seventeen 135
Appendix. 143
Afterword 146
What is Sericulture? 151
Index of People and Place Names 154

Foreword

Some 13 years ago, my wife was helping out in a Corowa op-shop when two strangers came in seeking information about Aitken, Gray and Neill, landowners in the mid 1800s and a mulberry farm at Corowa. Meeting them by chance that evening at the RSL club whilst having dinner, the question came up again. It was too much for me to resist, so a subsequent meeting was arranged. I informed them that there was an area just north of Corowa known as Mulberry Farm. It was a popular swimming spot for the locals over the hot summers, and I remember swimming there.

To me local history is always an interesting and fascinating place to visit, reminding us of what our forebears have achieved.

If one persists, information is out there somewhere – old photos, newspapers, written records and old buildings to visit etc.

Ian and Marilyn's researching and tabulating information on Mulberry Farm would have been most challenging and, may I say, rewarding for them.

As you will read, Sarah Bladen Neill was the major player in the development of the silk industry in Australia, with Corowa becoming an important part of the industry.

To succeed you must love what you are doing. Ian and

Marilyn's research and findings prove this point.

It has indeed been a great pleasure to help and show them around Corowa in their quest – researching the silk industry and the life of Sarah Bladen Neill.

Russell Black
Corowa, 2017

About the Authors

Ian Braybrook and Marilyn Bennet have lived and worked in Central Victoria for over forty years. Now retired, they were popular radio broadcasters and known for their outstanding community work. They are still active in the community and operate a small non-profit radio station for Castlemaine seniors.

Ian has previously written several books, each published and edited by Marilyn, but this is their first joint work on a manuscript. Both have a strong interest in Australian history and in its people.

Acknowledgements

Our thanks to:
The many people in Corowa who helped us so much in our research of their beautiful town; especially to Jan and Russell Black who have been treasures; the Federation Museum, the Corowa Historical Society, with help from Allan Handberg, Pat Deveson, Alan and Lesley New, the late Val Swasbrik, and their committees.
Also at Corowa – *The Corowa Free Press*; the late Dixie Lesley; Corowa Library; Cannon Bill Ginns and Rev Rex Everett from St Johns Church; Wendy, Margaret and Adrian from the Federation Council; Corowa Information Centre; the late Gus Peirce, author of "Knocking About" and Brian Burton author of "Flow Gently Past"; and at Rutherglen, Peter and Margaret Smith.
We are indebted to Len and Dyonne Rhodes for making us always welcome at the historic site and magnanerie on Cropper's Lagoon.
To Peter Kabaila, History Advisor to the Federation Shire, thanks for the interest in the site and the magnanerie and we hope you continue working to have this historic building registered and restored.
At Harcourt and Castlemaine area: The Harcourt Heritage & Tourist Centre who inspired the research initially – Lyn Allen, the late Yvonne Graham, George Milford and their

committees; Noel Davis; Heather Morrison; *Castlemaine Mail*; *Mount Alexander Mail*; Mount Alexander Shire; the former Metcalfe Shire; Castlemaine Historical Society; Castlemaine Art Gallery & Museum; Department of Environment, Land Water & Planning Victoria; Parks Victoria; Maree Edwards MLA and Jean Wyldbore. To our friends and family who have encouraged us along the way. Thanks Brook Acklom for the read and kind words.

To Bernard Schultz from Level Heading, thanks so much Bernie for your patience – your work is exceptional.

We must not forget:

The family of Jessie Grover, the Manager of Orbe Farm at Harcourt, Betty Grover and her late husband Harry, Hugh and the late Jill Smailes, and author Michael Cannon.

Jane Broadmore and her sister Susan Middleton, who are descendants of the Neill family – thanks for our day out at the Old Melbourne Cemetery and sharing the knowledge of your family.

Alan Telford, Ueili Ramsier from Germany and Professor Ponomir Tzenov from Poland, thanks for sharing your knowledge.

Research from the following sources has been invaluable:

State Library of Victoria; State Library of NSW; National Library and Trove; Public Records Office of Victoria; Victorian Heritage Register; East Melbourne Historical Society; Kew Historical Society; South Ayrshire Council Libraries; His Majesty's 40th Regiment at Foot re-enactment and history; Ancestry.com; Wikipedia.

To innumerable other newspapers and the many people of Corowa and Harcourt districts who offered help.

To any person or organisation we have overlooked, please accept our apology.

Prologue

*The Goddess of the Silkworm
A Chinese Folk Tale*

Hoangti was the emperor of China. He had a beautiful wife whose name was Si-ling. The emperor and his wife loved their people and always thought of their happiness.

In those days the Chinese people wore clothes made of skins. By and by animals grew scarce, and the people did not know what they should wear. The emperor and empress tried in vain to find some other way of clothing them.

One morning Hoangti and his wife were in the beautiful palace garden. They walked up and down, up and down, talking of their people.

Suddenly the emperor said, "Look at those worms on the mulberry trees, Si-ling. They seem to be spinning."

Si-ling looked, and sure enough, the worms were spinning. A long thread was coming from the mouth of each, and each little worm was winding this thread around its body.

Si-ling and the emperor stood still and watched the worms. "How wonderful!" said Si-ling.

The next morning Hoangti and the empress walked under the trees again. They found some worms still winding thread.

Others had already spun their cocoons and were fast asleep. In a few days all of the worms had spun cocoons.

"This is indeed a wonderful, wonderful thing!" said Si-ling. "Why, each worm has a thread on its body long enough to make a house for itself!"

Si-ling thought of this day after day. One morning as she and the emperor walked under the trees, she said, "I believe I could find a way to weave those long threads into cloth."

"But how could you unwind the threads?" asked the emperor.

"I'll find a way," Si-ling said. And she did; but she had to try many, many times.

She put the cocoons in a hot place, and the little sleepers soon died. Then the cocoons were thrown into boiling water to make the threads soft. After that the long threads could be easily unwound.

Now Si-ling had to think of something else; she had to find a way to weave the threads into cloth. After many trials, she made a loom – the first that was ever made. She taught others to weave, and soon hundreds of people were making cloth from the threads of the silkworm.

The people ever afterward called Si-ling "The Goddess of the Silkworm." And whenever the emperor walked with her in the garden, they liked to watch the silkworms spinning threads for the good of their people.

Project Gutenberg The Child's World Third Reader by Hetty S. Browne

The remains on Mount Alexander, about 1993.

Introduction

It was a hot February day in the second half of the nineteen eighties, deep in the pine forest on the eastern slope of Mount Alexander, where we two sat on a slab of granite. It was one of many roughly quarried slabs and pitchers scattered around the stone ruin that we had come to see.

Although the sun was still midway between its high point and its setting, the thick pine forest was bringing about early darkness. Dark or not, we found it hard to leave, somehow drawn to what we were experiencing.

We had only learned of this site on the mountain the previous day when on a field trip with people from the Department of Conservation and Natural Resources. They had taken several of us, members of the Harcourt Valley Heritage Centre, on a bus tour to point out various land systems in the area.

Here on the tree-clad mountain, our tour leader made an unscheduled stop at a large pine forest. He told us it was a school plantation planted in the 1950s as a fundraiser. He mentioned briefly that the ruins of a silkworm farm lay hidden somewhere within it.

A silkworm farm? Surely that can't be true. Here on this desolate, dry and rugged mountainside? It seemed very unlikely.

The roofless granite building we sat within was crumbling into disrepair but obviously it had once been a structure of substance, built to last. But for what purpose exactly? Other scatterings of granite pitchers, a few handmade bricks and traces of footings indicated that there may once have been other buildings. To add to the knowledge we were gaining of the area that we had made our home, we resolved to find the building's true history.

We didn't fully realise how difficult and complex it can be to delve into someone's activities 150 years on. Our lengthy research journey took us metaphorically to India, Italy, France, England, Scotland, Queensland, Tasmania, Western Australia and South Australia. We sought help from sericulture experts in France, Switzerland, Bulgaria and England.

We visited Corowa many times and the more we found out, the less we knew! In Corowa we received help from many people. This included farmer and horse breeder, Len Rhodes, his daughter Dyonne, and Allan Handberg from the Corowa Museum. Others were willing helpers, like Russell Black and his wife Jan who went up many blind alleys for us and produced welcome information.

We felt frustration, disappointment, annoyance, excitement and exhaustion, but in the end we were able to uncover this most remarkable story.

* *

Mount Alexander, 120 kilometres north-west of Melbourne, towers above the beautiful Harcourt Valley to the west and Sutton Grange to the east. It is a significant landmark. On 28 September 1836, Major Thomas Mitchell's exploration party, making their way homeward to Sydney in their travels through what Mitchell called *Australia Felix*, were the first white men to see it. As Mitchell waited for repairs to be completed on one of his wagons he went by horseback to the top of the mountain to take bearings. It rose an impressive 300 metres above the surrounding land and 740 metres above sea level. On a whim, the Major named it Mount Byng, after his friend Field Marshall John Byng, a leader of the British army during the Peninsular War. Not much later he changed it to Mount Alexander to honour Alexander the Great of Macedonia.

At that time the mountain was less timbered than it is today, in fact, from the summit, Mitchell was easily able to view the peak of what is now known as Mount Macedon.

Fifteen years after Mitchell's visit, the mountain was to give its name to the richest alluvial goldfield the world has ever seen, with nearby Forest Creek its centre. But that's another story.

In July 1845 Melbourne medico, Dr William Barker took up the land that is now the Harcourt Valley, 20 000 acres (8 000 hectares) in all, where he ran several thousand sheep. He and his brother Edward, also a surgeon, were well known and respected figures in Melbourne at the time. In 1852 he was appointed magistrate for the Castlemaine diggings. Interestingly, for many years it was believed that the site of the monster meeting of diggers on 15 December 1851 was on Barker's property.

Barker's land was resumed under the Duffy Act in 1862 and he returned to practice in East Melbourne. In 1863 he was appointed as surgeon to Beechworth hospital, then moved to Echuca two years later where he set up a practice. He remained there until returning to live and work at 43 Latrobe Street, East Melbourne in 1871. Both he and

Edward are remembered with the naming of Barkers Road in Kew for Edward and Barkers Creek and Barkers Streets in Harcourt and Castlemaine for William. In 1873, Dr Barker's wife, Madeline, appears as a shareholder of Sarah's Ladies Sericultural Company. William Barker features again in this story at a later time.

In future years the site of Barker's run became well known for its market gardens and apple and pear orchards, but that had to wait for a very important event, the discovery of gold on his property in September 1851.

Harcourt today is a small town of about 500 people, with a newsagency/general store, post office, motor garage/service station and a hotel that no longer operates. It is, however, earmarked for strong future growth due to its location alongside the Calder Freeway and close proximity to Bendigo, Castlemaine and ease of access to Melbourne.

In 1859 the construction of the Melbourne to Murray River railway commenced and large-scale quarrying of Mount Alexander granite began. This was for use on the numerous railway station buildings, staff housing, viaducts and bridges.

In 1864, that railway was to convey a huge sixteen tonne slab of the mountain's granite to Melbourne for the carving of the massive Burke and Wills monument. First erected in Elizabeth Street, it is now situated in the Old Melbourne Cemetery. At the same time, use of granite expanded to be a common building material, especially in Melbourne. In the Harcourt district a good number of homes were built from it. Many of these substantial buildings still stand and granite quarrying on the mountain continues. Blight's Quarry, which began operating at the time, is listed on the Victorian Heritage Register.

The original occupants of the mountain and surrounds

were the members of the Liarga Balug clan of the Dja dja Wurrung Aborigines. According to a census taken in 1841 they numbered forty in all. The clan was led by Munangabum, later named King Abraham by the white invaders. The natives had a different name for Mount Alexander. They called it Leanganook.

Whatever its name it was always going to be a difficult place to raise silkworms and grow mulberry trees for their feed. Who would be brave, ill informed or senseless enough to try?

Terms used in Sericulture

Grain – the egg laid by the silkworm moth from which the silkworm emerges.

Education – the development process from the hatching of the egg to the completion of the spinning of the cocoon and killing of the new moths – before their emergence from the cocoon.

Filateurs – winding houses where the cocoon threads are unwound, either by hand or machine. In the older days, by hand, but machines were available in the mid 1870s.

Magnanerie – as a farm is to agriculture so a magnanerie is to the production of silk. This is where the silkworms are raised to the cocoon stage.

Sericulture – the word describing the process of silk production.

This is the only known picture of Sarah. This is the gown she wore when presented to the Queen of Portugal. The gown was made from silk produced in Corowa and was adorned with green beetle wings.

Chapter One

I do not care
A single hair
For what some people say:
But this I know,
That silk will grow,
And I can make it pay.
Ann Timbrell, Collingwood, 1866.

Reports in the *Mount Alexander Mail* newspaper early in 1874 told of two women visiting Castlemaine to conduct a public meeting. The purpose of the meeting was to discuss the ladies' intention to establish a silkworm farm and 'school' for women on Mount Alexander, seven miles to the north. They were intent on offering a full explanation of their plans and encouraging investment in the scheme they were about to outline.

The visitors who alighted from the morning train were forty-five year old Mrs Sarah Florentia Bladen Neill and thirty-two year old Mrs Jessie Grover. Both ladies were impeccably dressed in the latest fashions and had an air of quality and good breeding. Indeed, Mrs Sarah Neill in particular was precisely of that background.

Sarah Neill was the widow of Colonel John Martin Bladen Neill, the highly regarded Deputy Adjutant General of Australia, Field Officer of the newly formed Volunteer Corps and Commander of the 40th Regiment, based in Melbourne.

Mrs Jessie Grover was born in 1843. She was a key person in the project about to be outlined at the meeting at the Mechanic's Institute. Jessie was Mrs Bladen Neill's right hand and her selection to be manager of the proposed Mount Alexander enterprise.

According to the Australian Dictionary of Biography entry by well known writer Michael Cannon, Jessie's great nephew, she was the youngest daughter of hotel keepers William and Elizabeth McGuire, who kept the Red Lion in Lonsdale Street, Melbourne. She had a deep understanding and commitment to sericulture, the process of silkworm farming and silk production, the subject they were in Castlemaine to promote. Aged thirty-two at the time, she was a bundle of energy and enthusiasm; as well, she was a close friend and confidante of Sarah. Of convict stock, her paternal grandfather Peter had been transported to Van Diemen's Land for highway robbery.

Jessie was one of four children, her father died in 1844 when she was just two. Her mother, Elizabeth, stayed on at the hotel but in 1848 she leased the business and instead ran a boarding house.

It was about then that Jessie was sent to live with her uncle Pat McGuire, at McGuire's Crossing. Pat kept the Emu Inn there and operated a punt across the Goulburn River. It is likely that it was there that she met, and later married, Harry Ehret Grover, a bit of an English toff. Harry was educated at Eton and came to Australia in 1851. Some say he was a remittance man, that is, one sent to the colony by his family having disgraced them somehow in Britain. It is possible, as it was not uncommon in those times.

Harry signed on as a police trooper, and when not on a gold escort, was based in the McGuire's Crossing settlement which later became Shepparton. He and Jessie married in 1869 at St James Church, Melbourne.

When Jessie's mother died in 1879 she left the couple well off, enabling them to buy a large home in fashionable St Kilda and to live happily on their investments. In the meantime, Jessie had met Sarah Florentia Neill and they became firm friends.

Sarah explained to Jessie the great passion she had for the potential of silk and silkworm production, and she became equally involved. There is more to the story of Jessie, Harry and their son Montague which will emerge later.

The meeting in Castlemaine went very well. Sarah explained the many benefits she saw in silkworm farming, correctly named either sericulture or sericiculture. In particular, she saw it as a potential employer of females, from children to the old. It was her determined intention to bring the ideas she had to fruition. Indeed, she had already begun sericulture work at Corowa in New South Wales and had recently formed the Victorian Ladies Sericultural Company Ltd. Jessie Grover had been appointed as honorary Managing Director.

The company had already obtained a government grant of land on the eastern slopes of Mount Alexander for the purpose of establishing such an industry. In all, the Minister for Lands had granted them one thousand acres. Why Sarah chose Mount Alexander is a mystery. Perhaps she believed that, based on her knowledge of European silk, high country provided the best environment for silkworms.

The future for Sarah and the industry was looking good.

At the conclusion of the meeting Sarah invited people to purchase shares in the company at four pounds each. A good number of those present were suitably convinced and made the purchase. Sarah and Jessie made it very clear however that, while men were welcome as shareholders, no directors of the company were to be male. It was a rule strictly adhered to.

Sarah and Jessie returned to Melbourne well satisfied with the support they had been given and proceeded to develop the site on the mountain. Their confidence was reinforced by a report they had commissioned that showed the land to be admirably suited to the purpose.

Prior to the visit to Castlemaine the company had engaged a person they believed to be a reliable, professional person to provide an assessment of the suitability of the Mount Alexander site. James Goggin, a civil engineer from Geelong, was chosen at a fee of one pound per day plus expenses and he was paid seven pounds in advance. In due course he provided a glowing report on the suitability of the land for the purpose of growing mulberries, the essential food for silkworms. Based on this report the project moved confidently forward. Alas, they had been badly deceived as they were soon to discover.

A report in *The Age* on 19 November 1873 had an interesting story to tell. Briefly, James Goggin had initiated court action against the Ladies Sericultural Company for non-payment of the fee balance owing. It backfired rather badly and it was shown that the ladies, not he, had been cheated. Evidence showed that the man concerned had never so much as visited the site and it was proven that he had completely manufactured the report from his office in Geelong. He was unable to tell the court where he left the train, in which direction the mountain was from Castlemaine, and was not able to give any description of the terrain. He was convicted of fraud and the judge awarded the Ladies Sericultural Company costs of three guineas. But the damage to the company's future project on Mount Alexander had been done.

Chapter Two

Meet Sarah and the Smith Neill family. The pride of the British Army. Death of Colonel Neill. Mr and Mrs Ralph. Baptisms and weddings.

In the early evening of 18 July 1859 a smartly uniformed military officer on horseback was making his way home on the Punt Road from the home of Captain Hans White of the 40th Regiment at Gardiners Creek (present day Kooyong), to his home in Kew. The rider's horse suddenly and violently shied and threw him to the road. The man lay motionless.

A short time later, at 7.30 pm, Patrick O'Shea, a local plasterer, walking to his home in Richmond, was passed by a riderless horse heading in the opposite direction. It was pursued by two barking dogs. Forty metres further on he sighted the form of a man in uniform lying prostrate on the road. He checked to see the condition of the motionless man and found him unconscious and bleeding from a wound to the head. At that precise time a cab was about to pass the scene so O'Shea hailed the driver for help. The cabby immediately drove the short distance to the Richmond police station.

Attracted by the sounds outside, two men emerged from the nearby Commercial Hotel to investigate. O'Shea at once sent one of them to urgently bring Dr McGregor who lived close by. He was on the scene quickly and hastily examined the man.

There was little he could do.

Two constables soon arrived and they picked up the still unconscious man and transported him in their cart to the police station.

The injured man was very well known and an important person so therefore was quickly identified as Colonel John Martin Bladen Neill. He was obviously in a very serious condition, bleeding severely from the ears. A messenger was hastily dispatched to bring his wife, Sarah, and she arrived at 10 pm.

Dr Alexander Gaul, a highly respected surgeon, was then called, arriving at 10.30. He too was unable to do much for Colonel Neill.

Despite every effort by the doctors, the Colonel died at 7.30 the next morning without regaining consciousness.

An inquest was held two days later on July 20. Dr McGregor testified that the injured man was senseless when he arrived at 8 pm but had muttered a few words about dogs, nothing more.

Captain White stated that he had felt that, although the Colonel had ridden the horse for two years, he considered it a "buck jumper that shied easily and was dangerous to ride". He added that when the Colonel had left his place at 6.30 he was entirely sober.

The reason for the accident was never proven although there was obviously speculation that the horse had been spooked by the two dogs seen by Patrick O'Shea. *The Argus* report of July 20 described the circumstances of the Colonel's death and noted that he had been on his way home after visiting fellow officer, Captain White. The article did not offer a suggestion as to the cause of the accident. However, the following edition stated that Colonel Neill's horse had been startled by a dog. Although probable, no actual evidence of this was evident,

Colonel Neill was the author of two books: "With the Somersets in Afghanistan : The Recollections of an Officer of H.M. 40th Regiment During the First Afghan War 1838–42" *and* "Recollections of Four Years' Service in the East with H.M. Fortieth Regiment". *Both are important records of the role of his regiment, (The Fortieth at Foot) in the first Afghan War. They include his regiment's involvement in the "Army of Retribution", which took them to Kabul where the British army had been annihilated. In retribution they left Ghuznee a smoking ruin, strewn with dead. Colonel Neill prepared his stories from the collection of letters he wrote to home in which he carefully recorded all significant events. Both books are valuable historical works giving insight into the war through to its conclusion. His books are still available.*

therefore the true cause is unknown.

The Coroner found that Colonel Neill's death was caused by a fractured skull

On 19 July 1859 Colonel John Martin Neill, the Deputy Adjutant General in the colony, Commander of the 40th Regiment and recently made commander of the new Volunteer Corps and, next to the Governor, Melbourne's most important and respected man, was dead.

As Deputy Adjutant General, Colonel Neill was the head of the British Army in Australia, the top man. His superior, the Adjutant General, Sir George Brown, was based in Britain in overall charge of the entire army and he was the man who appointed Neill as his deputy in Australia. He obviously held him in high regard. Therefore, Colonel Neill was a man of considerable power and inherited wealth. He began his army career at the bottom, having joined with the junior rank of ensign in 1833. He rose to his high position due to his outstanding military service.

Just three days before the Colonel's death, on July 15, the Governor, Sir Henry Barkly KCB, had issued two proclamations. He had used the provisions of the Volunteer Act of 1854 to raise this new defensive force. The proclamations specified the units be raised as ten companies of rifles and called upon "our loyal and faithful subjects in residence in Victoria to enrol themselves and be prepared to assemble for the purpose of drilling and instruction at time and place appointed".

This new volunteer force was to replace the small and separate units that had been established in Victoria at the time of the Eureka massacre in 1854 in which Colonel Neill's former regiment, the 40th, had a prominent role. Eureka occurred a year before the Colonel's arrival in the colony.

The volunteer force was scheduled to be officially proclaimed on July 27, a ceremony which proceeded under the heavy cloud of the Colonel's recent and sudden death. As an indication of the respect held for him, one of his 40th Regiment officers, Captain Jason McCarthy, named his newborn son "Bladen" in his memory.

Colonel John Martin Bladen had married Sarah in 1851. His bride was British born into wealth. He was from the highly respected and wealthy family, the Smith Neills of Barnweill and Swindridge, extensive properties near Ayr, Scotland. Ayr is an historic seaside town with sandy beaches, with about 50 000 people. It is situated on the west coast of Scotland and its most famous inhabitant was poet Robert Burns. There is a statue of him in the town. The Neill's family tree is very impressive, dating back to the 17th century. Originally the name was simply Neill, until an ancestor, William Neill assumed the name Smith. He did this in return for his inheritance of the large Swindridge property from his wife's uncle, George Smith, which he added to the Barnweill property. William became William Smith Neill and for a generation that family name was retained.

Sarah's husband John Martin Bladen Neill, had two brothers, the eldest being Brigadier General James George Neill. A professional soldier, he was declared a British Military hero for, among other things, his role in crushing the mutiny against the Empire at Lucknow in India.

The History of the Landed Gentry in Great Britain and Ireland, volume 2, says,

> "He was a soldier whose name will live in history, whose services in India and the Crimea were of prominent importance. At Lucknow, here at the point of a bayonet, on 25th September 1857, at the

very moment of complete success, fell the gallant General Neill, the pride and idol of the army. The queen at once raised his widow to the rank which he was about to receive, Commandership of The Bath."

She also granted his widow, Isabella Neill (née Warde), a pension of a significant five hundred pounds a year. General Neill is honoured by an impressive statue in Wellington Square, Ayr and a replica thereof in Madras, India. The tiny thirteen square kilometre island of Neill, in the Ritchie archipelago, governed by India, is named in his honour.

The story of the siege of Lucknow, when Indian troops of the British East India Company rebelled, is an epic in itself. Since the time of the rebellion opinion is divided on the heroic role of General James George Neill due to the horrific nature of punishment handed to the Indian troops.

The General's father, Colonel William Smith Neill, had found time from his war battles to marry Caroline Skiller and father three sons – Sarah's husband John Martin Bladen Neill, Captain William Francis Smith Neill and Brigadier General James George Neill. The latter, when not off fighting wars, fathered nine children, the youngest son being James John Vansittart Neill. He was born in Madras, India, which indicates that the General took his wife with him to that place while he was fighting battles. We will hear more of young James John Vansittart Neill later.

Brother William also died in military service of Britain, at Colombo, Ceylon (now Sri Lanka) on 26 September 1862. Thus the three sons of William and Caroline Smith Neill died in war service. Can we imagine the grief of the mother, Caroline?

On the death of her son John in 1859 she wrote to a Mr

Campbell at Ayr who had organised the erection of the statue of her eldest son, James. There had been a meeting of town dignitaries where the latest Neill death was discussed. The meeting asked Mr Campbell to convey the condolences of those present. Mrs Neill wrote a rather disjointed response:

> "My dear Mr Campbell. In response to a letter of condolence wishes of the meeting brought together of those who inaugurate the statue raised by the country to the memory of my eldest son General Neill, to express to me their deep sympathy for this added calamity, to yourself for proposing it, and to Lord Eglinton for his most feeling acquiescence, and the other gentlemen at the gathering, I beg to offer my most grateful thanks. God knows I need something to alleviate the bitterness I feel."

John Martin Bladen Neill and Sarah Ralph married twice. The first service for the Scottish side of the family was held on 6 June 1851 at Dalry, near Ayr. The second was in St Paul's in Middlesex, England on 12 June 1851, six days later.

Both would have been extravagant and very formal affairs, due to the families' social standing. Why the two weddings? It appears that wedding number one was for the Neill family in Ayr and number two for the Ralph family in London, not uncommon then, as now.

The Ralph family lived on South Molton Street, in the Mayfair area. South Molton Street ran from Oxford to Brook Street and today is famed as a street of high fashion.

Sarah was born there in December 1828 and baptised Sarah Florentia Ralph on December 12 in the magnificent St George Church. This historic building remains on Hanover Square, just a few blocks from her home.

Sarah's father, Edward Ralph, and mother, Everel, were from the society set of the time. Edward was described as a "gentleman" which meant he was of the upper level of society.

We have little knowledge of John and Sarah's day-to-day life leading up to their marriage, nor of how they met. Of course, we know of John's continuing involvement in the military service but nothing of Sarah. Probably, as with all ladies of the time, she led a very sheltered life in the family home until a proper suitor emerged. Obviously they would have moved in the same level of society.

A little over a year after the marriage, on 10 July 1855, Colonel Neill (then Major Neill) was listed to embark from Kingstown, Dublin, for Melbourne with the members of the 40th Regiment at Foot, under the command of Colonel T.J. Valiant. Major John Martin Bladen Neill was the officer in charge of the troops. They consisted of nine officers, eleven sergeants, eight corporals and three hundred and fifty-six privates. Major Neill was paid an additional three shillings a day for the voyage. They sailed on July 15 and arrived at Hobsons Bay, Melbourne a little over three months later on October 24. Their ship was *HMS Vulcan*, as named on Hart's List of members of the 40th Regiment at Foot.

The regiment and passengers didn't disembark until November 5, held on board because of an outbreak of typhus fever, a serious infectious disease, on the recently arrived migrant ship *Ticonderoga*. One hundred passengers had died and almost everyone on board was sick with one thing or another. The *Lysander*, a ship of 500 tons, had been hired for the purpose of accommodating the 40th Regiment. However it was instead diverted to serve as a hospital ship and anchored alongside the *Ticonderoga* and fitted out to take fifty beds. The 40th were forced to wait on board the *Vulcan* until the typhus victims had been moved to quarantine at Point Nepean before moving to barracks ashore.

No evidence has been found, but presumably Sarah accompanied John on the ship, as many other wives and children did so.

Always one to be involved in the community, within weeks of his arrival, John Martin Bladen Neill joined the newly established "Nightingale Fund". This organisation was officially posted in the *Government Gazette* as "In support of the training, sustenance and protection of nurses and hospital attendants." The group included the cream of Melbourne society with names like Justice Barry and John Pascoe Fawkner prominent. Sarah and Colonel John Martin Bladen Neill had no children.

On arrival, Major Neill was officially advised of his expected promotion to Deputy Adjutant General with the rank of Colonel. It's interesting to note that Edward McArthur, the son of John McArthur of merino sheep fame, was the Deputy Adjutant General from whom John took over. McArthur was married to John Martin Bladen Neill's sister (also named Sarah).

***

Colonel Neill's funeral was one of the biggest ever seen in Melbourne, rivalled only by that of Governor Charles Latrobe. It is fair to say that it was a spectacular event. The double coffin, coloured black, was set atop a gun carriage drawn by four fine horses adorned with silver trappings and plumes. It was draped in a Union Jack on top of which was his sword, hat and belt. The cortège journeyed from Apsley Place Military Headquarters, where he had lain in state, to the New Cemetery. It extended for about two kilometres, with every vantage point occupied by spectators. The graveside service was conducted by Bishop Perry, watched by some three thousand people. Almost every dignitary in the colony was present, including Governor Sir Henry Barkly, judges, barristers, politicians and councilors.

The double coffin, one within the other, was lowered with due reverence into the ten-foot deep grave, itself fully lined with bricks. Three hundred soldiers of the 12th and 40th Regiments stood to attention, with rifles placed muzzle down until given the order to fire three rounds as the coffin passed from view. According to a report in *The Age* of July 20 the volley from Neill's Regiment, the 40th, actually dislodged a male spectator from his perch in a tree. He hit the ground hard but was not badly injured. The incident caused much suppressed mirth among the crowd of onlookers.

In half an hour the scene was deserted apart from several workmen whose job it was to fill in the grave and cover it in concrete. Next door stood the imposing monument to the late Governor Charles Hotham.

Chapter Three

Sarah inherits Corowa interests and moves there. Rivalry between Sarah and Ann Timbrell. Government reluctance to accept sericulture. St John the Evangelist Church. Andrew Hume. Corowa – Federation conference, 1893.

With the passing of her husband, Sarah became a wealthy woman. She not only inherited their home in fashionable Kew and a considerable sum of money but there were large interests in a significant number of properties in and around Corowa on the Murray River in New South Wales. John had previously entered into partnership with two established property owners in the district, J.H. Aitken and George Gray. Sarah inherited the one-third share held by the late John.

In the Supreme Court of New South Wales.
ECCLESIASTICAL JURISDICTION.
In the will of John Martin Bladen Neill, late of Turnville Kew in the Colony of Victoria, a Lieutenant-Colonel in Her Majesty's Service, deceased.

NOTICE is hereby given, that Sarah Florentia Neill, of Corowa, Murray River, in the Colony of New South Wales, widow, the sole Executrix named in the will of the above testator, intends, at the expiration of fourteen days from the publication hereof, to apply to this Honorable Court, in its Ecclesiastical Jurisdiction, that probate of the said will, or of an exemplification thereof, may be granted to her, as such Executrix as aforesaid.—Dated this 3rd day of September, 1868.
WILLIAM GEORGE PENNINGTON,
Proctor for the Executrix,
108, Elizabeth-street, Sydney.
4269   6s. 6d.

The partnership included the Brocklesby property, which was later to be made famous by the Tom Roberts painting done there, "Shearing the Rams". Of interest is that in approximately 1866 the property was offered for sale by the partnership with 3500 head of cattle – quite a large stock number. Some of the smaller properties adjoined the Murray River east of the current Corowa saleyards on what is now Honour Avenue. It is close to town and it is there that Sarah Neill planted her first ever experimental mulberry trees in 1865.

Being a woman of independent spirit, it was natural for the widowed Sarah to be attracted to Corowa where she now shared ownership of so much property. She made the move to the town within two years of the death of John, an area that she loved and almost always referred to it as her home "on the river" rather than Corowa. Although a more

Turinville, Kew, as it was in the 1850s. This was the home of Colonel and Sarah Neill at the time of his death. (Courtesy NGV.)

> TO BE LET.—That delightfully situated residence, with the grounds, on the banks of the Yarra, known as
>
> TURINVILLE, KEW.
>
> The house, which has been recently painted, papered, and put into complete repair, consists of drawing-room, dining-room, breakfast, parlour, three bed-rooms, bath-room, kitchen, with range, &c., servants' room, pantry, and store-room, and two rooms adjoining, suitable for bed-rooms. The out-offices are very complete, good stables and coach-house, men's huts, &c. The water is laid on from the Yarra to the house, the stables, and the garden.
>
> The flower and kitchen gardens are in good order, and the orchard is well stocked with fruit trees in full bearing.
>
> The farm and grounds adjoining are extensive, and may be let either with the house or separately.
>
> The present proprietor and occupier, J. R. Cowell, Esq., being about to proceed to Europe, will, on letting the premises, cause to be disposed of by auction the whole of his elegant and useful household furniture, thus giving the in-coming tenant an opportunity of purchasing any part thereof.
>
> For further particulars and cards to view apply to Dr TIERNEY, Registrar's Office, Prince's Bridge, Melbourne. 4ʊ 58 4 8

or less permanent resident of Corowa Sarah maintained a substantial home in Barkly Terrace, a group of six residences in Grey Street, East Melbourne – a prestigious address. The building has long since been demolished and now is a car park – a reflection of modern times. It was about this time that Sarah adopted the name Sarah BLADEN Neill; probably it had a better ring to it, more suited to the society in which she moved!

It is not known what first excited Sarah's passion for sericulture, perhaps it was her love of fine clothing and items made from pure silk. Maybe she knew of it from England. In any event, excited she certainly became. Most likely she had read a paper published by J.F. Nerevy of Richmond in 1861 titled "The Silkworm – Silk is Gold". This would certainly have captured her interest.

In his paper, which was comprehensive and well written, Nerevy illustrated the potential for the industry and outlined the pitfalls. He gave details of the costs and benefits and gave examples of the openings for the industry in Australia. For example, in 1850 England had imported an astonishing fifteen million pounds worth of silk. Even France, which had produced six million pounds worth of silk, had imported two million pounds worth. This was big money back in 1850. As many others before and since, he emphasized the potential for farmers and small cottage industry for families. The entire cost of education equipment for a cottager was about ten pounds. There was also the need for mulberry leaves, a recurring problem given the huge appetites of silkworms. Another disincentive was the long wait for a return on the mulberry trees; about 4 or 5 years before proper harvest of leaves was possible.

Over the years between 1860 and 1890 the constant theme was, for the industry to be established from zero, government assistance was needed. Successive governments showed a distinct lack of interest. This quite likely denied Australia what would probably have been a successful industry in sericulture.

An interesting item appeared in *The Farm and Station* on 3 August 1868. In it the writer lamented the lack of progress in silkworm farming and highlighted the most significant problem as lack of tree planting to provide food for the worms. The delay of four to five years on any return was the principal reason for farmers and investors shying away. On the plus side, there was the very low cost of entering the industry. "No branch of rural industry," proclaimed the writer, "showed more potential for profit." Still there was little reaction and the industry remained largely ignored.

The article continued, "The small cost includes the big advantage of minimum building. Trees were easily propagated at almost no cost and a thousand trees would provide income equal to one thousand sheep. A further advantage is that the work occupies only about three months of a year."

Freedom from disease was also raised as a serious requirement and one which was not easy to accomplish, as demonstrated in Europe. Once again the need for government financial and other support to establish the industry was raised.

* *

After John's death, with no family in Australia and being without children, Sarah probably sought something to keep her interested and occupied. Having read so many published articles concerning the farming of silkworms, it's little wonder she turned her attention to that industry. Among the papers she read was no doubt the beautifully presented, hand-written and illustrated pamphlet published by Ann Timbrell of Collingwood in 1860. Ann Timbrell was undoubtedly the pioneer of sericulture in Australia. Sadly, after a lifetime of endeavour, she died in poverty in the Old Colonists Home in the mid 1890s. These homes were established by the renowned actor and philanthropist George Coppin in 1870. It was a place for colonists who had arrived in the colony before 1851, who had fallen on hard times. Coppin began by donating the first small bluestone cottage of the 144 or so currently on site. It remains today. He also established the Victorian Humane Society and St John Ambulance Service. The Old Colonists

Homes are still in existence in the Melbourne inner suburb Fitzroy and are listed on the Victorian Heritage Register. Sarah Neill always acknowledged the role and support of Ann Timbrell, but there was a strong rivalry between them.

Ann Timbrell's interest began as a child in Yorkshire, England when, as did many children of the time, she grew and bred silkworms as a hobby. When she settled in Melbourne in 1860 she began her sericulture venture. The next year she joined the Acclimatisation Society which had recently been established by Edward Wilson, retired editor of *The Argus* newspaper. A prominent member of the group was the famed Baron Von Mueller, founder of the Melbourne Botanical Gardens

The Society's purpose was to introduce new species to the colony and preserve those already in existence.

Unfortunately, although he did much great work, Baron Von Mueller's exotic imports such as the mulberry also included the blackberry and gorse, which quickly became

The site of the silk enterprise of industry pioneer Charles Brady, on the Tweed River at Tumbulgum, New South Wales, 1881.

pests and noxious weeds.

It is interesting to note that the magnificent Melbourne Zoo was also established by this wonderful organisation that Edward Wilson headed.

At that time Ann Timbrell made contact with the well known sericulture enthusiast Charles Brady, who was not yet setting up his sericulture farm at Tumbulgum on the Tweed River in New South Wales. Brady, a lecturer in agriculture at Hawkesbury College, who still lived and experimented at Curl Curl, eighteen kilometres to the north of Sydney, happily supplied her with thirteen cocoons free of charge from his healthy stock. This involved the valuable cargo in a lengthy journey by train. At the same time he offered free grain (i.e. eggs) to anyone else in the colony interested in developing a sericulture farm. Mrs Timbrell and Mrs Neill were two of few who had accepted.

In line with the two women, Brady also saw sericulture as a promising industry and estimated returns of ten pounds per acre, a huge figure in that era. He sought government help in his work and in putting in place teaching of the industry in schools for future employment. Both requests were firmly rejected. No government interest in sericulture appears to have existed, but the New South Wales government later did give some help with land. Their failure to recognise the potential for sericulture set the standard for all governments, excepting perhaps Western Australia. It was government short-sightedness in the extreme.

Baron Von Mueller happily assisted Mrs Timbrell in her sericulture efforts by allowing her to harvest mulberry leaves from his infant tree plantings in the Melbourne Botanical Gardens and the Kings Domain. Her husband Andrew was unable to help much, being at work each day in his job as a shorthand writer, presumably in a court. Two or three times every day Ann Timbrell walked from her newly established North Collingwood home, rain or shine, to gather hessian sacks filled with leaves, essential to feeding her worms; they would eat no other and had huge appetites. The creatures taste in mulberry leaves included several varieties, but their preference was for black mulberry and the alba.

Meanwhile, Sarah Neill began her work at Corowa and was making quite a name for herself. Perhaps because of this there was a rivalry developing between the two women.

In one of Ann's published papers she writes about the adverse effect of heat on the silkworms. "How does this section apply to Mrs Colonel Neill's proposal of a system of 'open air'? Has that Lady any idea of our hot winds and dust storms?"

On another page, "Mrs. Neill complains of losing 30,000 worms, but she has fed them with damp leaves. Surely she should have known that this cannot be done."

Since 1865 Ann Timbrell had sold 27 pounds of her cocoons. By 1867, on her new farm on Plenty Road, she had grown almost 400 mulberry trees, having begun with a handful of saplings. By propagation from cuttings the number had quickly grown, as had her cocoon production. In 1872 she sent a batch of cocoons to an agent, W.C. Middleton, in Brighton, England. He replied that he had sent them on to Italy as they did not wind silk in the United Kingdom. However he described her product as "absolutely first rate". Three hanks of silk resulted and Ann entered them in the London and Vienna Exhibitions of 1873 where she was a winner. She was one of the most highly regarded sericulturalists in the Colonies.

One of Sarah's first jobs in her new home town of Corowa was to erect a suitable monument to her late husband, John.

She made contact with many of her wealthy friends both in Australia and the United Kingdom and from them, heavily subsidised by herself, she oversaw the building of the original St John the Evangelist Church. The beautiful weatherboard church was completed about 1863. This building was eventually demolished and replaced with the present substantial brick building in 1911. Within its walls is a plaque taken from the original building honouring the memory of John Martin Bladen Neill.

Corowa, with a population of about 5600 lies 280 kilometres north of Melbourne and 610 kilometres south of Sydney, is a pleasant and prosperous town on the northern bank of the Murray River. The famed Brocklesby run, which formed the basis of the town, was established in 1839 by Charles Cropper, his name is perpetuated in "Cropper's Lagoon", which is prominent in this story. Cropper later took in a partner named Thompson. By 1841 Tom Chapman was leasing the property and he eventually sold to Hamilton Hume, the famed explorer. Hume took it up for his sister-in-law, Elizabeth, who was widowed when her husband John was shot dead by bushrangers at Gunning. Her son Andrew managed the property but sadly he died in 1847 aged just 31, to become the first person buried in the Corowa cemetery.

Oddly enough, when first surveyed, Corowa was in a different location to today. A map of 1861 shows it to be where the current South Corowa is situated, approximately where the racecourse and golf club lie. Nobody seems to be sure why the development ignored the original plan, perhaps it was fear of flooding, but in any event it appears to be the logical site. The bridge at the base of Sanger Street, Corowa's main thoroughfare, connects nearby Wahgunyah in Victoria, and has probably contributed to the town's changed development.

Corowa has an important place in Australia's history. The town had a notable role in the establishment of the Federation of Australian states, mainly due to the town being closely involved in the payment of tolls inflicted on interstate trade at the time. The Corowa Conference of 1893 is forever recorded in the story of Federation. Corowa is now the centre of the Federation Shire which resulted from an amalgamation in 2016. Across the river is the old river port town of Wahgunyah, in the days of the riverboats, a very busy port.

Corowa shot to fame in 1951 when the deadly rabbit disease, myxomatosis, was detected killing rabbits in abundance along the river town flats. This followed numerous experiments, designed to introduce the South American disease, which appeared to end in failure. The success at Corowa came from the millions of dusk-biting mosquitoes inhabiting the weedy lagoons. Rabbits, bitten by the insects, died in millions. It was not known until then that it required mosquitoes to spread the disease – and Corowa had them to spare!

Chapter Four

Mulberry Farm. The Europe and United Kingdom visit. Attention to mulberry trees and silkworms. Death by disease in Europe.

In 1865 Sarah made her first planting of mulberry trees on one of her small pieces of land just outside Corowa. With stock obtained from Charles Brady from the Tweed River area, most likely with the assistance of the "Chinamen" she hired from time to time, she planted over 200 seedlings of cape mulberry. This was more as an experiment than anything else. Without watering systems the cuttings were left to their own devices as she left on a journey to Europe early in the next year. She had decided not to spare herself, or her money, in her efforts to establish a silk industry in the colony, convinced that it had a bright future. She was particularly dedicated to the ideal that sericulture would provide employment for females and additional income for farmers. In a pamphlet she wrote at the time she spoke of her concern for farmers' uncertainty of crops.

In April 1873 in an interview with the *Beechworth Ovens and Murray Advertiser* she said,

> "I turned my attention to Mulberry trees and silkworms. If a farmer could supplement their income with silkworms it would surely assist them. In the culture of silk is to be found the

> solution of the problem as to how agriculture is to be made pay in the colonies. The growth of grain and the production of silk by the farmer's wives and daughters, in my opinion, will enable the farmers to tide over bad seasons, and to hold onto their crops in good seasons until the markets prove remunerative."

She firmly believed that silkworm farming would fit well with the farmers' work and, with family assistance, provide a buffer if their crops failed, as so often happened. She also stated that she had first thought of silk culture around 1865.

> "There was an abundance of land and it occurred to me that something ought to be done to turn that land to profitable account. I thought of something that would pay and my attention had been turned to silk. I made contact with Sir James Ferguson and Dr. Black and obtained my first cuttings and samples of Japanese grain. I had at Corowa tested to see how far the culture of silk would succeed. In the first year the Japanese yielded a lot of small but good cocoons but in year two they were diseased. I then determined to go to Europe to find a cure for the disease."

So in 1867 Sarah set off on one of her several visits to the United Kingdom and Europe, specifically to see if a disease cure existed and to learn all she could firsthand about sericulture. Embarking on a long journey by sailing ship to Britain at that time was not an easy thing to do; her courage in travelling alone was remarkable, a reflection of her dogged determination. This visit to Europe was relatively brief, seeking as much expert advice as was available. She was shocked and disappointed to find that the silk industry in Europe was in serious decline due to the rampant pebrine disease, for which there had been found no cure.

Pebrine is the common name for a difficult to control protozoal disease. In the 19th century it was almost impossible to control. According to silkworm pathologists it is caused by strains of microsporidia. The worms are infected by transovarian and horizontal transmission. Affected worms become flaccid and fail to grow. Death results. Today it can be controlled by destroying all moths after they lay their eggs and the body ground to paste. Under micro-examination the microbe is revealed and any diseased batch of worms is immediately burnt.

While in the United Kingdom, Sarah renewed contact with old friends, many of them from the hierarchy of the time; people she knew through her marriage into the wealthy Smith Neill family of Barnweill and Swindridge. Her list of friends is astonishing and reads like the *Who's Who* of Britain. One example is Viscountess Strangford, society writer, illustrator and nurse, a darling of British society. As with many wealthy women of that age the Viscountess had done valuable work in the local community as well as overseas. She had worked in community health in Bulgaria and had been a volunteer nurse at University College Hospital, London. She had recently been awarded the Royal Red Cross by Queen Victoria for setting up a hospital in Cairo. She died from a stroke in 1887 on board the *Lusitania* journeying to Port Said to establish a hospital there. The Duke of Manchester introduced Sarah to the Princess of Wales who, some time later, ordered Australian silk hose for her children from her.

Sir Antonio Brady was another important person who was

eventually to assist Sarah. He was the son of a storekeeper and began as a junior civil servant and went on to become Superintendent of the Admiralty. He was knighted by Queen Victoria in 1870. In his retirement he focused on geology and natural history, which included an interest in sericulture. He was also the brother of Charles Brady, the leading sericulture authority and promoter in Australia who greatly assisted Sarah.

Others were Lady Lucy Calvert, daughter of the Earl of Powis, George Montague, Lady Mills, Lady Neville, Lewis and Allenby of London, Silk Mercers to the Queen, and famed English silk manufacturer James Brocklehurst MP. These and many others were soon to become shareholders in Sarah's eventual bold sericulture venture in Australia.

Her influential contacts were many, extending to royalty, and she didn't hesitate to utilise them in her sericulture endeavours. The same group of friends also subscribed to the building of the Colonel John Bladen Neill Memorial Church, St John the Evangelist, in Corowa, which Sarah described as "the handsome little sacred edifice".

Laden with information and bursting with ideas to expand her dream, Sarah arrived back at Corowa in the late summer of 1867. She was astonished to see that her 200 small mulberry trees had thrived and grown considerably. This in spite of being completely unattended and enduring what had been a very hot summer. They were all heavy with an abundance of leaves; silkworm food. This delighted her no end and further encouraged her to continue with her work

Chapter Five

Sarah starts work. Good eggs! Ann Timbrell. Monsieur Roland, Charles Brady, Reverend Docker, Bontharambo and more.

On her return from overseas Sarah began her work in earnest. Armed with greater knowledge she embarked on what we call today a promotional tour. She sought publicity wherever it became available, spoke at town meetings and wrote lengthy letters to newspapers. All had the same theme; the sericulture industry was absolutely right for the colonies, for farmers and for investors.

She travelled extensively in the days of stagecoach and the horse and buggy, setting up town meetings all across southern New South Wales and north-east Victoria. She became a familiar sight and extremely well known. It seems that nothing could stop her in her relentless pursuit of acceptance of sericulture as a worthwhile industry.

On one occasion she had addressed an evening meeting at Benalla and had booked a hall in Wangaratta for the following night. She was already aware that the Reverend Joseph Docker of Bontharambo, a large property a short distance from Wangaratta, was interested in the industry. In fact he already had a thriving experimental three acres of well established mulberry trees on his land. Sarah had made the acquaintance of the family due to her nephew James having courted, and later married, the Reverend Docker's niece, Alice Bristow. James and Alice were married in the magnificent homestead in 1874.

It being holiday time in the region Sarah found to her dismay great difficulty in renting a horse and buggy in Benalla. It appeared that she may not be able to get to Wangaratta in time for the meeting so she telegraphed ahead stating it unlikely she would make it. However, she moved heaven and earth and finally was able to hire a horse and buggy privately. Driving frantically, she made it to Wangaratta right on time, exhausted but not at all flustered. The attendance at the meeting was poor due to her telegraph message but she was not bothered. In her friendly and conversational style she addressed those assembled for almost an hour and a half; cajoling, appealing to greed, persuading, whatever it took. Sarah was adept at public speaking. As a result, a group was formed to begin a sericulture project, utilising the tree plantation established by Reverend Docker.

She continued her work at Mulberry Farm, cultivating mulberry trees and experimenting with raising silkworms. She supported the development of a sericulture farm by the Beechworth Sericulture Company by giving generously of advice and supplying them with mulberry cuttings and seedlings, later providing them with quality grain at moderate cost. Eventually the Beechworth project was to fail and there were recriminations as attempts to lay blame were placed on Sarah for providing faulty grain. This she vehemently rejected.

1869 was a red letter year for Sarah as she harvested her first Mulberry Farm cocoons, although limited in number, describing them as "small but good". Importantly, she had proved to herself and any doubters that she could do it.

Throughout the year and into 1870 she obtained a quantity of good Japanese grain from her friend and mentor, Charles Brady from Curl Curl, New South Wales and from the Acclimatisation Society in Melbourne.

At that time Mrs Ann Timbrell, by then a leading member of the Acclimatisation Society, had leased 15 acres of land at the three mile post on Plenty Road, Collingwood, where she continued to expand her sericulture venture. She named it Timbrell's Silk Farm.

Mrs Timbrell went on to exhibit her silk in Dublin and followed that at the London International Exhibition, where she was awarded a medal for quality product. In 1877 she was awarded a medal for "Merit in Sericulture" from the Acclimatisation Society for her work. This medal was sold privately in 2015 for $1000. Regrettably her venture did not succeed over time as she had hoped.

Many people were very willing to help Sarah Neill, this often came from her contacts in high places. As an example, in 1870 the South Australian Botanic Gardens, under orders from Sir James Ferguson, the governor of South Australia, delivered to her an assortment of mulberry seedlings. Dr Black, President of the Acclimatisation Society and Charles Brady each presented her with quantities of good grain of several varieties.

By 1874 Mulberry Farm had a plantation of an impressive thirteen acres planted with 12 000 fast growing mulberry trees. They were so closely planted as to be sufficient for 100 acres, said one visitor. The many cutting frames on the grounds contained about 23 000 new cuttings, 15 000 being cape mulberry and 8 000 alba (white) that she had imported from Italy and France.

Sarah had also settled on using the "open air" system, as

recommended by the Swiss expert Monsieur Alfred Roland, and had built a magnanerie to his specifications, about one hundred feet long and twenty feet wide. She also added a small leaf room for preparation of the leaves for the silkworms' food. Her little farm was becoming a big attraction in Corowa.

At around this time, with her workload expanding, she employed two labourers. There was much heavy work. She also appointed Miss Helen Stuart, presumably a local girl, as her principle assistant at Corowa. It proved to be a good choice.

Chapter Six

*Off to the continent – again. The pebrine
disease. Louis Pasteur's treatment. Sarah
swears to give up silkworms and silk forever.
Antonio Brady. Under Roland's wing.*

Sarah was restless, anxious to learn more from the experienced experts from Europe. She also wanted good quality grain for Mulberry Farm at Corowa. So, in 1871, she boarded a steamer and again set sail for Italy, taking with her samples of her cocoons.

Her first call was Naples where she was horrified to learn that there, as in most of Italy, sericulture farmers were giving up. Thousands of mulberry trees had been dug from the ground and been replaced with livestock or potato crops. Returns from the Japanese grain available were too small to be profitable. The dreadful disease, pebrine, was rampant and seemed unstoppable. Sericulture was simply no longer profitable.

Sarah realized that if she could overcome the disease and produce good grain in Australia it would be brilliantly successful. She envisaged a large export industry, involving not silk, but the grain produced by healthy worms.

Next stop was Florence and then on to Rome where she heard the same stories. Some said they were getting better results on the slopes of the Apennine Mountains where more space and air seemed to create a healthier environment.

She went on to Milan where it was a similar situation, with the few remaining farmers crying out for good quality grain. She visited a number of magnaneries and filateurs (reeling establishments) and everywhere found that good quality silk was an exception. Even Novi and Genoa, both famed for the quality of their silk, revealed no good grain. Sarah said later, "This certainly cost me a year but the knowledge I gained and the relationships I formed proved ample compensation."

Moving on, Sarah visited Nice, Grasse and Marseilles where she made further good contacts as well as gaining letters of introduction to people who shared her passion. She was told that in Spain and Portugal they were suffering from the disease and in every direction mulberry trees were being ripped from the ground. "Next year," she said, "I hope to visit Spain for myself and see just what is happening there."

At the city of Lyons the industry talk was of the value or otherwise of the "Pasteur system". Many believed it to be a successful method of controlling the disease, but it perhaps lacked sufficient support from producers as the disease continued unabated. At Lyons her showing of cocoons from Corowa impressed many but such was the despair and non-belief in the industry's future that there were no takers.

Meanwhile, efforts to control the disease were making progress. In response to a request from the government, scientist Louis Pasteur had begun working on a solution to the deadly sickness in 1865. (In France the disease was known as pebrine, French for pepper, due to the appearance of a pepper-like substance on the body of sick silkworms.)

Pasteur eventually isolated two diseases, not one, as was always believed. He discovered that it all began with diseased eggs and told farmers that these must be destroyed immediately they hatched. After a series of well publicised failures over several years, he and his team finally identified the cause as diseased globules carried by the parent moth. The globules proved to be alive – a microbe. Pasteur also found that healthy worms became sick and died after eating mulberry leaves contaminated with droppings from sick worms. This produced the second disease which had confounded him and thwarted his work. His solution was absolute cleanliness, but it appears that this may have been difficult to convey to farmers, and the disease continued. Pasteur provided the reason and preventative measures, but not a cure.

Badly disillusioned, believing she could never succeed under the circumstances, Sarah returned to Milan to stay a month with friends, very unhappy. She said later, on her return to Australia, "I made up my mind then to give up grain and silk forever."

She began preparing for a visit to England to visit family and friends. However fate intervened and she was the recipient of a stroke of good fortune.

Visiting a bank to arrange conversion of money to pounds sterling she met with the bank manager. She happened to glance at a small bag which lay on his desk. To her interest and astonishment it contained what was obviously a number of silkworm cocoons. Sarah told the man of her interest in sericulture, the purpose of her visiting Europe, and of her disappointment and problems.

> "All my plans were immediately changed. He told me he was a good friend of Monsieur Alfred Roland of Orbe in Switzerland. Hope was revived and it was through his kindness that I was introduced to Monsieur Roland. 'Madam,' he said, 'you will find what you seek in Switzerland.'"

Monsieur Roland was the leading expert in silk production and silkworm diseases; renowned for producing splendid grain and silk.

Sarah cancelled her trip to England and, bearing a letter of introduction, she instead undertook a long and often difficult journey to Orbe. Today it is a short drive of about five hours from Milan to Orbe but in 1871 it was a long, long way over rough terrain, across mountains and over streams, with roads not much more than muddy tracks.

> "Monsieur Roland received me most kindly and I was quickly installed in a house belonging to his friends. He seemed pleased that I had come so far to see his 'open air' system."

Sarah stayed for four months!

The good Monsieur Roland took her under his wing, assuring her that he believed that Australia's climate, combined with his methods, would bring success. He gave her open access to his facilities and every possible assistance. This included a promise to provide her with some of his precious, healthy grain.

She observed his reeling facilities and was given a suggested method for his open air system. As well as his magnaneries, Roland was experimenting with growing and harvesting the education outdoors, on actual mulberry trees. In her explanation of the system to the public, Sarah made plain the meaning of "open air" system. "Open air means breeding in a freely ventilated building, all doors and windows of gauze. Strictly, all walls should be gauze. It should keep out insects. Small vent holes in the roof covered in gauze are to be placed where no moisture can enter." It was this method that Sarah took back to Corowa.

The excited Sarah returned to Lyons where she met with merchants to ascertain the prospects for her if she could supply quality cocoons. She then returned to Switzerland and Orbe to further discuss her plans for 1873 and 1874 with Alfred Roland and to collect the promised grain. Sarah then travelled to Paris and London, spending a month in England and Scotland visiting family and friends. She also purchased a variety of items for the construction of a magnanerie in Corowa.

Having obtained the grain from Alfred Roland, it was essential for it to be kept in a controlled atmosphere. In particular, it would be necessary to keep the grain as cool as possible as they entered the warmer southern hemisphere. The idea was to fool the grain into believing they were still in the cold European climate and prepare them for emergence when they felt the warmth of Corowa. Her friend Sir Antonio Brady came to the rescue. He was well acquainted with the owners of the P&O Line and arranged for the company to construct a special icebox for Sarah. Their engineers built an insulated box that, if filled with fresh ice every day, would maintain a constant temperature of 44 to 50 degrees Fahrenheit (10 to 12 °C). The required amount of ice was loaded on board ship as part of the service! Brady also arranged free passage on the ship home.

Before going home Sarah was persuaded to return to Milan to see a new type of mechanical reeling machine. She wrote, "It is a slow reeler that any intelligent colonial girl could learn to operate in a few days." This brilliant invention was a huge improvement on the type of reelers in general use and Sarah stored this information for future use. Incidentally, the wage for a "colonial girl" at the time was twenty-five pounds a year!

Chapter Seven

*The Victorian Ladies Sericulture Company.
Influential membership. Growing more trees.
Wherever grows the vine, the silk is sure to thrive.
Publicity and promotion.*

Sarah's plans showed steady progress in 1873. Most importantly she moved to establish the Victorian Ladies Sericultural Company Limited, a bold and unprecedented act. The Memorandum of Association stated its objectives as – "the promotion of sericulture in the Australian Colonies by planting and cultivation of such kinds of species of mulberry trees as found to be the most beneficial food for silkworms." The rules continued with "obtaining grain, its education and reproduction, the production and reeling of silk."

The liability of the company was limited, with capital of five thousand pounds divided into 1250 shares of four pounds each. The memorandum then listed the original shareholders, approximately fifteen in total, all female; the number one shareholder's signature being Sarah Florentia B Neill.

A report, lodged soon after with the Registrar General by the newly appointed honorary Managing Director, Jessie Grover, showed the membership had expanded to 563. The registered address of the Company was number 36 William Street, Melbourne, but by 1875 it had moved to Goldsbrough and Co. Wool Store in Bourke Street, Melbourne, and then to 47 Market Street.

The 563 membership list of 1873 looks like the *Who's Who* of society, mostly female; wives of leading squatters, doctors, members of Parliament, solicitors, judges, surveyors, Knights of the Realm, sea captains, bank executives and reverend gentlemen. Standing out was one male exception at that time, Frederick Standish, Police Commissioner, whose residential address was The Melbourne Club! Standish, of course, featured prominently in the pursuit of Ned Kelly several years later. Another was famed actor of the period, George Coppin. Some distinguished names were Madeline Barker, wife of Dr Edward Barker, John DePass, a wealthy merchant, Ann Timbrell, sericulturalist, Theresa Casey spouse of the Minister for Land, Frances Hackett, wife of Judge Hackett, and Elizabeth Docker of Bontherambo, whose husband, Joseph, has appeared in this story. Interestingly, the 1875 return shows Jessie Grover's address as Harcourt.

* *

With the company up and running Sarah Neill's focus was now on developing her operations and growing more trees on Mulberry Farm. She had settled on using the open air system as recommended by the Swiss expert Alfred Roland and had built a magnanerie to his specifications. She also added a leaf room for preparation of the leaves for the silkworms' food.

As well as Mulberry Farm, in 1874 she had planted about 1600 mulberry trees on the seventy-four acre piece of land the Gray, Aitken and Neill partnership owned just below Croppers Lagoon. She was preparing for the need to have many trees to supply her expected huge number of hungry silkworms. She had also planted twenty-two acres of grape vines on that land around 1873, apparently considering a venture into wine making. Perhaps she had heeded the words of Olivier de Serres, the father of French agriculture: "Wherever grows the vine, the silk is sure to thrive."

She also recruited other nearby landholders to grow mulberry trees as food for her worms.

The result of her venture into grape growing is not recorded. The only mention she ever made was at a meeting where she suggested that she had had an experience in wine making that she preferred to forget.

She now began a campaign publicising the potential sericulture industry and developing the ladies' company by putting into effect the plan she had had since the beginning. She and the company would promote the industry through every available event.

Agricultural shows were common across the colonies, giving opportunities to display the product and speak about it. She resolved to speak there and at public meetings in towns and townships wherever she could find an audience. She would involve farms and cottages.

Charles Brady and Monsieur Roland would supply her with good quality grain and she would appoint grainers to distribute it to growers. The grainer would train the new recruits. Each grower would receive fifty grains free of charge, complete with proper instructions and advice from the fully trained grainers. Every grower would share the profits from the resulting grain sales as follows: one third to the grainer, one third to the grower and one third to the partners, Brady, Neill and Roland. This, she believed, would stimulate interest and growth in the industry.

The Victorian Ladies Sericulture Company would sell the product and be a distribution centre for grain and profits.

The grain would be sold in Europe and estimated initial production was five thousand ounces valued at one pound per ounce – five thousand pounds, a considerable sum in 1873. She would train as many women as possible in the business of sericulture. The Victorian Ladies Sericultural Company would do the same.

Sarah now proceeded with the part of her plan to divide the colonies into Sericulture Districts, which brings us to Mount Alexander and Castlemaine.

Chapter Eight

Castlemaine. Jessie and Harry Grover, managers. Progress on the mountain. A dispute with the Metcalfe Council. Bad eggs to Beechworth? A costly undertaking.

Following the meeting at the Mechanics Hall in Castlemaine early in 1874 Sarah and Jessie returned to Melbourne. Sarah was to remain there to oversee the conduct of the company and, until a replacement was found, to be the teacher at the new Sericulture School for Ladies in Grey Street, East Melbourne. Jessie was soon to be transferred with her husband to manage the branch on Mount Alexander.

Another sericulture school, with several pupils enrolled, had been operating at Mulberry Farm in Corowa for several weeks. It was in the hands of the trusted Miss Helen Stuart, whom Sarah had hired in 1871 and who had since gained much experience.

With cheery and confident goodbyes Sarah boarded the train from Spencer Street to Wodonga. The journey home to Corowa was much easier in 1874, following the opening of the new North Eastern Railway to Wodonga the previous year. Alighting at Chiltern it was a relatively easy journey by coach to Corowa.

The project at Mount Alexander was not all smooth sailing; far from it. To begin with, access to the site was far from easy, placed as it was high on a mountain slope in rough and timbered country. Transporting building materials and necessary stock, including furniture, tools and so on was a nightmare.

It was far from being a suitable site for sericulture, but the land was free and the two women were acting on what they believed was expert advice from the Geelong based civil engineer. Hadn't Monsieur Roland himself advised Sarah that high mountain country was admirably suited for raising and educating silkworms. In fact he had been referring to alpine country with fresh air and regular snow and ice. Although it can be very cold and quite frosty, snow and ice are a rarity on Mount Alexander. Had Sarah misunderstood him? It is possible and it would give a further explanation for her proceeding with the unsuitable mountain farm.

As well, there was the matter of cost. *The Victorian Farmer's Journal*, in an article in the mid 1860s, detailed some of the costs likely to be involved. It emphasized that the high price of labour should not be a deterrent to sericulture farming. The writer firmly believed that the "high labour charges" would be overcome and would not absorb the profit to any detrimental extent. The writer went on that if the disastrous spread of disease as in Europe could be avoided, profits in the colony would be quite good. He also pointed out that in Europe, the farmers, most of whom had regular farming to attend to, used the downtime in off seasons to work on the sericulture part of the business; time that otherwise was not productive. His arguments made good sense.

The comparative costs that he estimated would have applied approximately to Sarah and the Ladies Sericultural Company.

Calculated to suit the Company's twenty acre undertakings

the figures are as follows:

Purchase 20 acres of land @ £2 per acre	£40
Preparing land for planting @ £20 per acre	£400
Planting 20 acres of mulberry	£1600
Two labourers for one year	£116
Women's labour for 4 years @ £41 per year	£164
Extra labour and expenses	£190
Buildings etc.	about £1000
Total	£3510

This would have represented a considerable outlay for the company, even though the supply of trees and grain was contributed *gratis* by Sarah Neill.

In the verbose writing style of the day, the writer concluded positively:

> "It would appear that silk culture has nothing to fear by contrast, either with the price of labour in Victoria, or the competition from foreign produce; and that on every consideration, it would form one of the most useful adjuncts to the agricultural productions of the colony."

On 27 July 1874 the *Mount Alexander Mail*, an enthusiastic supporter of Mrs Neill and the Grovers, reported on the progress of the project on Mount Alexander.

> "Some thousands of mulberry trees are planted which have made such progress as to be bursting into leaf. A house had been begun, as yet unfinished, but habitable. In a short time it will be completed and the school can be opened for

instruction of those ladies who desire to be taught the art of sericulture."

There was no mention of the difficulty any such pupils would have in accessing the site. The track to the school was in very bad condition, so bad as to make access very difficult and according to the *Mount Alexander Mail*, unsafe. The ladies had told their reporter of the difficulty they were having, not only with transporting construction equipment to the site, but in allowing visitors. In the delightful writing style of the time the article went on.

"There is extreme difficulty to the approach to the Farm and something needs to be done to it quickly. When this is done the School will form a pretty visiting spot for those who, in summer, enjoy a ramble among the wild and vastness of Mount Alexander. That this road should at once be put right is the bounden duty of the Council – it would be too great a tax on the philanthropy of the spirited promoters to undertake on their own responsibility."

Unhappily, the Metcalfe Shire Councilors, who were mostly district farmers, were not so supportive. In fact they were quite hostile toward the ladies and their "innovation". Why, some asked, did a woman secure a grant of one thousand acres, especially of land grazed free by the farmers of the Metcalfe Council. There was strong opposition and resentment among some men of the time. This is reflected in the attitude adopted by the men of that Council. There was a belief that no woman or women were worthy of a government grant of one thousand acres of land to grow mulberry trees and feed silkworms.

An example of the sexist attitude is shown in the following extracts from a poem, "Haunted by Worms", published in *Melbourne Punch* on 22 July 1875.

It tells the story of fictitious Bob Billydad who aspires to election to the Metcalfe Shire Council.

> - Being haunted forever by ghosts of silkworms,
> He'll try to get cancelled the Ladies's Reserve –
> The Silk farm – a grant that they didn't deserve.
> ...
> "These silkworms," he said are most dangerous brutes,
> And positive foes to all other pursuits;
> They break down our fences, they eat up our grass,
> They cut down our timber, and sell it, alas!
> 'Tis unsafe to travel the bush too, you know,
> Through the worms of the Sericultural Co.–
> ...
> "I'll teach all these 'girl of the period' belles
> To carefully mind of what Bob Billydad tells;
> I'll teach them the law about feminine rights!
> They shan't wear no breeches – I ought to say 'tights.'
> Ah! Could I but punish those growers of silk,
> I'd curtail their *trains* and cremate 'em, like Dilke;
> ...
> For daring to come to this district to plant
> A mulberry farm on their Government grant.

There was indeed, serious objection to the farm occupying part of the Mount Alexander Common. As reported in the *Mount Alexander Mail*. "The Mount Alexander common is huge. Thousands of acres, most of it rich flats and pastures. There is an abundance of good land." Indeed there was

Jessie Grover

Harry Ehret Grover

plenty of land and it is difficult to accept that a sericulture farm that occupied barely twenty acres would have any impact whatsoever on the farmers grazing cattle on the mountain. Apparently the farmers, who had grazed the mountain for many years, felt ownership of it. They also objected to giving access to the farm through their land. Entry to the farm was on a rough path from the south-east side of the mount. To use the path required the opening of slip rails at three points along the way, which apparently greatly annoyed the farmers whose land the path crossed.

Access to the farm was also available by the bridle path that led from what is now the Harcourt Oak Forest in Picnic Gully, to the mountain top. A bridle path is suited only for traffic on foot or by horse.

* *

Toward the end of 1873 Harry and Jessie Grover, with son Montague in tow, had taken up residence at the farm and all was in readiness to begin the teaching of young ladies. Two had been selected from the group at the Domain and they would share the small home with the Grovers. The description of the farm as a "school" sometimes used in the press was probably derived from a misunderstanding of the term "education", used in the silk industry to describe the process from hatching to harvesting the cocoons.

As well as the cottage, a magnanerie, a copy of what she had seen in Europe, had been built and a large leaf storage shed completed. The *Mount Alexander Mail* of 17 November 1874 described the scene and painted an excellent picture, it was obviously a farm in full operation.

"A very pretty cottage has been erected for Mr

> and Mrs Grover, the lady manageress of the establishment, and the pupils; this is situated at the extreme south end of the allotment and is surrounded by out-houses suitable for a farm. About 100 yards west is erected a very commodious building called a magnanerie and in this the feeding and rearing of silkworms is carried on. The upper part of the ground is trenched, and planted with white mulberry trees and cuttings of various varieties, amongst which might be mentioned the 'rubra' from North American ... French ... and Chinese ... and ordinary Cape."

In reality the "pupils" were workers undergoing training, who rose at dawn and ceased work at dark, every day of the week. In total some 33 000 worms were housed, all under cover to protect them from the frost.

The following description of the magnanerie given by a contemporary writer is practically the same as the one Sarah built at Mulberry Farm but different to what she later built at the Croppers Lagoon site.

> "The magnanerie is an extensive building, being 92 feet long by a breadth of 29 feet. The windows, of which there are five on each side, are not glazed but consist of openings fitted with zinc shutters and fine crinoline muslin which freely admits air and light but excludes intruding insects. The ceiling is of the same material. The roof is of shingles. The zinc shutters can be closed in the event of inclement weather. The suspended frames, six in all, contain four layers of eight trays

in which the worms at all stages can eat happily of the mulberry leaves provided.

"A new building has been erected called a leaf room. In this are stored the leaves to be sorted and prepared and it will soon be used as a rearing room, too."

The writer continued with considerable enthusiasm.

"The energy displayed by Mrs. Neill and Mrs. Grover is deserving of every encouragement and will surely tend to establish the new industry in our midst, one that will be heard of throughout the world."

In January 1874 an important visitor alighted from the train at Castlemaine, bound for Mount Alexander. It was none other than Charles Brady himself, bearing a supply of healthy grain he had bred at his establishment on the Tweed River. He met at the farm with Sarah, Jessie and Harry, expressing his delight with what he saw and offered firsthand advice. This, of course, was readily accepted. Brady was acknowledged as the foremost authority on sericulture in Australia and we see again the extent of Sarah's circle of friends.

A further report appeared in a December 1874 edition of the local newspaper. The reporter wrote: "In the course of the next few weeks the insects will begin their education, namely spinning their cocoons". However his next visit to the farm in the autumn of 1875 resulted in disappointment: "It was discovered that owing to the severe frost experienced the proper supply of food had been destroyed and none was to be procured. The managers had been compelled to stop the operation."

Mrs Grover soon obtained a new supply of eggs from Corowa and with the mulberry trees quickly growing new leaves and those gathered from friends, all was in readiness for another education.

The *Mail* at this time reported on the condition of the grain that had been lost and rumoured to be faulty and diseased.

> "It was part of the same parcel from which the Beechworth Sericulture Company was supplied by Mrs. Neill and Mr. Brady and no disease existed among them. At Harcourt every grain (egg) had hatched, a circumstance that does not always occur – the seed was magnificent".

The *Ovens and Murray Advertiser* had recently run an article suggesting that diseased grain had been sold by Mrs Neill and Mr Brady to the Beechworth Sericulture Company which had caused the collapse of the business. The *Mail* concluded:

> "It was not founded on fact. There were four other places where the grain was sold and at each of them the same success as at Mount Alexander. The cocoons from which education Mrs. Grover is taking to the exhibition in Melbourne, will be taken home (England) by Mrs. Neill."

Charles Brady had actually delivered the grain concerned to Jessie and Harry Grover at the mountain site, giving detailed instructions on managing the education process.

The article went on to describe how to gain entry to the farm, one way was via the bridle path from Harcourt and the other from the Sutton Grange–Chewton Road "by way of Mr. Ellery's and Mr. Beaumont's properties". This track had been promised by the Shire of Metcalfe for some months but they had been slow in acting. As it was, it presented a very difficult job to navigate.

After passing through three sets of slip rails the traveller was confronted with – according to *The Mail* – "a track strewn with boulders and trees, too narrow for any vehicle. An attempt to study the convenience of the visitors or proprietors is simply abortive. Something will have to be done. The least the Council could do would be to put it in passable order", thundered the writer.

For some months Mrs Grover had been attempting to have the road made passable by the shire council and for several months the shire had procrastinated. Letters flowed back and forth as Jessie and the council locked horns over the costs and need for the road. Indeed the Council appeared to be the opposite of supportive of the entire venture and

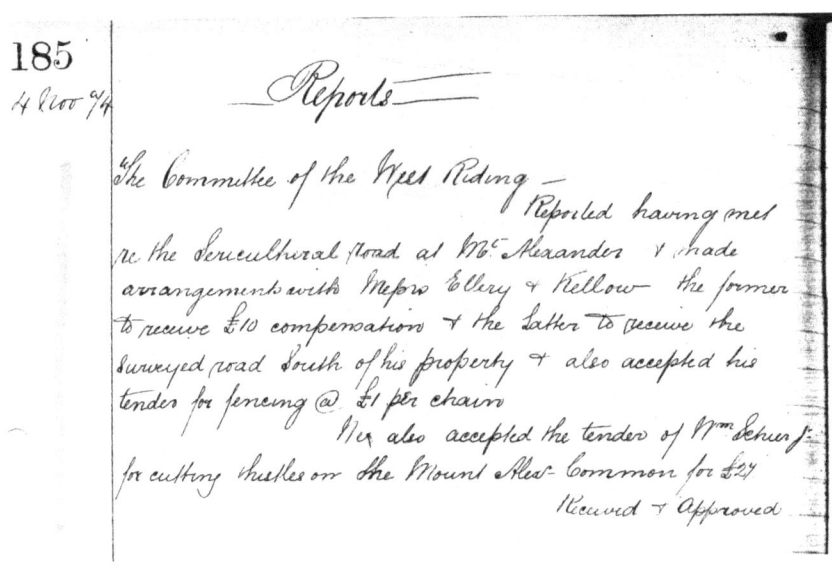

A report from the Shire of Metcalfe on the development of the road to the new sericulture farm.

was definitely not cooperative. Remember that the council consisted almost exclusively of farmers, many of them utilising the free grazing land provided on Mount Alexander. The government classified the mountain as part of the Sutton Grange common, available for grazing. In those days it was not as heavily timbered as today and offered extensive areas of grassland.

The letters began on 28 June 1874 and concluded in April the next year. In August the council finally agreed to fix the road providing the cost was no more than fifty pounds and the

A letter from Jessie Grover to the Metcalfe Shire.

Association was prepared to share this equally. The Secretary of the Company, Ms F.E. Tripp, replied that the Directors had agreed to the demand. By September no progress was apparent so Mrs Grover wrote a stinging letter to the shire asking that they put a stop to the expense and inconvenience required in the carriage of goods to the farm.

By November the council meeting decided that they would pay compensation to farmers Ellery and Kellow for access to the sericulture site via their land; ten pounds to Ellery and free construction of a road into Kellow's property. They also accepted Kellow's tender of sixty-nine pounds for fencing.

In December the council wrote to Jessie advising of their decisions at "an expense to this council of over sixty-nine pounds to fence the road and fifty-four pounds in compensation for taking part of the land (presumably Kellow's and Ellery's) and sixteen pounds for road given in lieu with deeds thereof."

In response, Jessie wrote that the Company was prepared to share the costs but declined making payment until the work was completed. She also objected to the high cost of fencing saying that "the Directors have been informed that the work can be done at a much lower rate."

In February 1875 the directors agreed to pay the promised half share of twenty-five pounds as agreed previously. No further payment was sought by the council and finally the job was completed. No trace of this road remains today. One of the letters to the shire secretary bore the address, "Orbe, Mount Alexander". It seems that either Sarah or Jessie, most likely Sarah, had named the property after the town in Switzerland where she had stayed with Monsieur Alfred Roland.

Chapter Nine

*Government House. The Domain.
A new Orbe on the mountain. The Governor
and the Minister pay a visit. Gifts from the
Ministers. The death of many worms.
An invitation to Mr Brady. Fifteen pupils.
Two thousand trees to Western Australia.*

While the argument with the Metcalfe Shire was going on Sarah had been very busy elsewhere. Having prepared a good quantity of grain ready for the Mount Alexander farm, she was faced with a difficult situation. The grain was going to be ready for hatching in the first education on the mount, but because of the roadwork delays, and problems in transporting goods to the site, Orbe at Mount Alexander was not ready to receive them. Sarah expected the further delay to be short, but as Corowa was far removed from Mount Alexander, there was a need to store the grain at a more convenient location. It is a measure of the power of her contacts in high places that Governor Bowen granted her permission to erect a temporary magnanerie on the Domain, part of the Government House grounds. Not only that, she was given use of space in Government House. This was unheard of for any ordinary citizen.

At the time she wrote in a letter to the *Town and Country Journal*: "The government has granted the Association 1000 acres of land in a locality suited in aspect and soil for a silk farm." (The original grant on Mount Alexander was in two pieces, in total 1 000 acres. Sarah had advised the government that 1000 acres was necessary to grow sufficient mulberry trees for production on the large scale that she planned.)

On the subject of the arrangements to receive the grain she wrote:

> "In order that this seasons should not be lost a small portion of land in Government Domain has been lent for immediate use, in which many valuable mulberry trees are already planted and the Company has procured and planted several thousand cuttings, so that in this important particular, no time has been lost. A room in the new Government House has been placed at the disposal of the Directors."

Sarah went on to say that she confidently expected the government to come forward with pecuniary assistance, noting that her friend, the Governor of Western Australia, hoped that two thousand pounds would be in their estimates for the next year, to assist the industry in that colony.

The Argus on 5 May 1874 reported on the activities in the Domain.

> "The Victorian Ladies' Sericulture Company is about to vacate the premises in the Domain lent to it by the government, and to settle down on the land at Mount Alexander, near Castlemaine. Preparations are being made for the move accordingly. The graining season will end in about a week and consequently, instead of having

to carry away worms and moths the Association will only have to pack eggs, a much easier undertaking".

Another indication of Sarah's strong links with the colony's hierarchy is demonstrated when in May 1874, the Governor of Victoria, Sir George F Bowen, the Governor of Western Australia, Mr James Weld, and the Minister for Lands visited her at the Domain before she left for Corowa. Governor Weld expressed his desire to establish the silk industry in his colony and had travelled to Melbourne expressly to learn more.

The Argus commented:

> "His Excellency, Sir George, expressed great surprise at the wonders accomplished by these lady pioneers of the new industry. About a half dozen young ladies have been admitted as students into the establishment and they do most of the work. The occupation is more likely to produce a sound mind and body than dawdling up and down The Block or lying on a sofa at home."

It seems that nothing much was different in 1874 to today concerning young people!

**

How the contemporary newspapers managed to obtain and cheekily publish private letters we cannot explain. Under our current privacy laws it could certainly not happen today. The following extract is from a letter published in the *Rockhampton Bulletin* of 5 June 1874. Sarah wrote to her business partner and friend Charles Brady mentioning her good relationship with government ministers. In it she made this surprising statement:

> "Our Ministry are [sic] doing us much good. Mr. Francis and Mr. Casey have given us three houses; and we all go to work. Botanical gardens give us all their old Mulberries; we must plant in hedges until we can cultivate more ground."

In the same letter she explains that she has fifteen pupils and other people's pupils plus the Ladies Sericultural Company to provide with grain from September to January. It will be a busy time for her, her student girls and supervisor Helen Stuart in Corowa at Mulberry Farm.

Her pupils hail from Adelaide, Corowa, Western Australia,

Example of a rearing house. This one was in South Yarra.

New Zealand, Tasmania and Sydney. In response to a request from the Governor of Western Australia, who had previously sent a girl student, Sarah announces that she is dispatching 2 000 mulberry trees to him on the next available steamer. She also invites Charles Brady to visit her at Corowa for a week, saying of her home: "It is as curious as myself and you will find it all very rough, but jolly. Never mind, we will come out before long and astonish not only the colonies but all of Europe. We should go home (England) together". It seems there were no personal secrets kept from the newspaper readers of 1874, but no doubt "coming out" had a different meaning to today!

The years 1873 and 1874 were indeed busy for Sarah Neill. In late January 1873 she spoke at a meeting she called in Melbourne, in the rooms of the Bank of New South Wales, a meeting attended by sixty-five interested people. She was, of course, seeking the participation of as many as she could muster to help in establishing the industry. She advised the meeting that she had established a training magnanerie at her home at number 6 Barkly Terrace, East Melbourne for the purpose of training a limited number of people in the trade. Anyone interested was welcome to attend.

She spoke at some length about the Roland "open air" system where the life of the insect was managed outdoors, as much as possible on the mulberry trees themselves. She said.

> "The system has received opposition and non-acceptance, particularly in France, where it was not understood and because it interfered with the old time honoured usages. His system is the nearer approach to natural and is therefore better than more artificial methods."

She added that Monsieur Roland believed that it was

desirable to establish in cold mountain areas where the grain could pass through the hibernation period at proper temperatures. Was this the reason for the choice of Mount Alexander in 1874? Speaking confidently, as she always did, she assured the audience that once a few ladies had taken up the teaching of daughters and wives of farmers, others would follow in their wake.

> "It will soon look like a flock of sheep following one another into a yard. Seven ladies have already taken up the charge of districts in which to teach the business. I first tried five years ago. The first trial failed from disease which was through my lack of knowledge and I made up my mind to go to Europe and learn about the industry and I did so."

Sarah Neill had no doubt that her plans for the growth of sericulture would succeed.

In February 1873 she was at a meeting in Albury urging her audience to take up the business. She emphasized the profits to be made, the excellent future prospects for sale of good quality grain in Europe, the employment of women and girls and the advantages for farmers who often had to take poor prices for their produce. As a result of this talk a silkworm farm was established on the outskirts of Albury by partners Thomas Affleck, a solicitor, and John Howard. They had soon planted 2000 mulberry trees which did very well and promised a good return. Their property, Glenmoris, was about three miles from town, consisting of ten acres. Within a year the number of trees totalled 11000. This was a serious attempt to establish the industry. It was supervised by Miss Affleck, one of Sarah Neill's pupils. The magnanerie contained no method of heating and therefore the temperature sometimes fell to as low as fifty degrees Fahrenheit (10 °C). At other times it increased to ninety degrees Fahrenheit (32 °C), not a good environment for silkworms. In addition, the worms were fed whole leaves, instead of them being cut into pieces as Sarah instructed; labour saving but definitely not desirable. It did not bode well for the future.

The entire stock was lost to disease in 1877 and again in 1878. The project was then abandoned.

Wherever she went, Sarah's talks all had the same theme. Several days before this event she had again visited Wangaratta. Here she described the process involved in raising silkworms, producing grain and the education process. Her energy and enthusiasm, impressed audiences wherever she went.

The *Ovens and Murray Advertiser* reported on her visits.

> "Mrs. Neill stated that the crops the farmers cultivated did not pay; the wine growing trade did not pay. She said she spoke from previous experience on that score having some time ago attempted to raise grapes and produce a small quantity of wine. She blamed the vignerons for the failure because they rushed immature wine onto the market."*

Sarah advised those interested to apply to the government for a grant of a small tract of land in the neighbourhood of the town, confident it would be provided. She also urged those present to consider taking shares in the Victorian Ladies Sericultural Company. She said that the shares were fixed at

---

* In the authors' investigations we found no evidence of her actually producing wine at Corowa, although she had planted twenty-seven acres of vines near Croppers Lagoon.

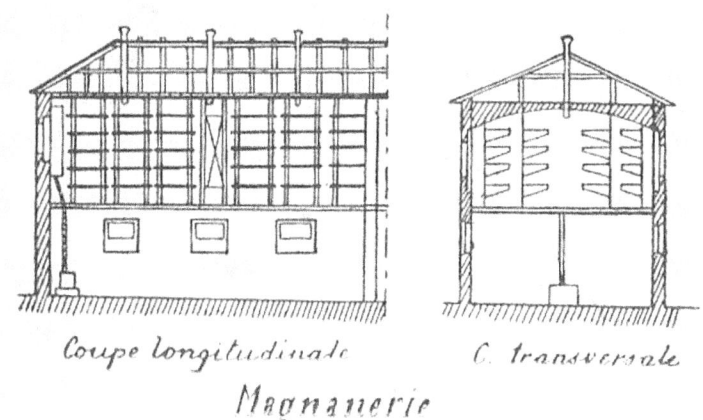

Plans for a magnanerie.

a very moderate amount of four pounds and from these she felt assured that a return of ten or twelve pounds would be reaped in a couple of year's time. An acre of mulberry trees would be enough to educate four ounces of grain. Each ounce would produce one hundred ounces; a ready market was available in Europe at one pound per ounce.

> "Mrs. Neill explained that she believed that the Cape Mulberry was the most suitable food for the silkworms and that anywhere vines would grow, mulberry would prosper. She had with her a variety of leaves to show, pointing out which were desirable and which to avoid. She offered assistance with the supply of plants and cuttings and concluded the meeting by again showing the advantages to be derived by farmers and free selectors from engaging in sericulture."

Sarah was a natural salesperson and in true style she urged her audience to take action at once to establish the industry. The implication was that to delay would mean missing the boat, a proven sales tool. The response she received was excellent and the foundations of several branches of the association were easily laid. As a result of these meetings several branches in north-east Victoria eventuated and thousands of mulberry trees were planted. Sarah's dream of a huge industry was becoming a reality

In January 1873 Sarah received a shipment of twenty mulberry trees and about 80 established cuttings direct from a plantation in Venice. The precious new tree variety arrived in excellent condition considering their long voyage. They were promptly and carefully planted at Mulberry Farm by well-known nurseryman Jack Cole. He said that the trees were healthy and each five feet high; the layered cuttings were also very strong.

With a huge number of worms to be fed in November of that year, an urgent appeal was issued in the local press for assistance in the supply of leaves for their food. Mrs Neill told the paper.

> "In three weeks time it will be over. Hundreds of worms are now spinning cocoons with thousands yet to follow. The cocoons are larger and as fine as the best specimens I have brought from Europe. But there are thousands of worms who must die unless contributions are obtained. All of the coach proprietors around Corowa have been most liberal in providing free carriage of leaves".

She was hopeful that the now opened North Eastern Railway which was within fifteen miles of Corowa would supply her with the necessary food.

> "I am thinking that the railway will give free

carriage for leaves as well. Address the bundles to me, Mrs. Neill, care of the Traffic Manager and I shall receive them. Removing the leaves from the mulberry and the under leaves will not hurt the tree; on the contrary it improves them".

The situation was dire and led to the deaths of many thousands of otherwise healthy worms, a setback to be overcome.

Chapter Ten

*Governor, Sir George Bowen visits Mulberry
Farm for dinner. A dusty problem.
The company is doing well.
Sarah selling lots of trees.
'Candaha', a Corowa magnanerie, described.*

In 1874 things were moving ahead at Corowa. A report in the *Hamilton Spectator* reprinted in *The Argus* of October 23 describes another visit to Sarah by the governor of the colony, Sir George Bowen. This was a huge event for the districts of Chiltern and Corowa. The governor was accompanied by the Attorney General, Mr Alfred Kerferd, Mr Witt the Minister for Lands, and several ladies from the Ladies Sericultural Company. The new North Eastern Railway from Melbourne to Wodonga was completed in 1873 making the journey to Corowa so much easier. Chiltern was the nearest station to Corowa and it was there the official party alighted from their special train. From there they would travel by a special coach, drawn by four horses, to Wahgunyah and thence Corowa. The party was entertained by an extravagant luncheon in the Chiltern town hall where a number of equally extravagant speeches were made before the party moved off to Wahgunyah.

The journey to Wahgunyah was seventeen miles over bumpy roads, no doubt to the discomfort of the royal's representatives. At the hamlet of Christmastown the entire population of eleven turned out to wave. After a short stop at Wahgunyah, it was off to Mulberry Farm on the Albury Road at Corowa where Sarah entertained the group with dinner.

In the evening Sir George chaired a lecture by Judge Augustus Fellows. Sarah was a special guest. The group stayed in a hotel overnight before making the journey back to Melbourne.

Visits by this impressive and important gentleman were rare and usually confined to ceremonial occasions involving the "aristocracy". Again, he expressed his admiration for Sarah's work and her plans in glowing terms.

Mulberry Farm had become a landmark in Corowa and remains so to this day. The site is still referred to by older locals as Mulberry Farm. As one unnamed old timer we met in the street said to us, "I remember it when I was a boy. A cottage with a corrugated iron roof and two big sheds. We always called it Mulberry Farm".

The *South Australian Register* of 30 June 1874 reported with considerable confidence:

> "With the proper supply of food and of grain there is practically no limit to the extent that the industry may be profitably pursued.
> "A report obtained by Mrs. Neill about cocoons she sent to Venice a few months since, submitted to judges, were found to be in every case their excellent quality and marketable value have been commented upon. They were declared of the quality of the product from Italy in the best days, before the disease."

The writer continued in the flowery style of the day:

"The results obtained at Mrs Neill's farm, Corowa, and at the Ladies magnanerie in the Government Domain are referred to as being particularly gratifying. The quality of the cocoons and the grain produced have been pronounced by Chevalier Marinucci, the Italian Consul at Melbourne, to be admirable."

Meanwhile, two serious problems were arising at Mulberry Farm; the large amount of dust being deposited on the mulberry leaves due to an increase in traffic on the Albury Road, and possums eating the precious leaves. However, Sarah was very pleased with herself. She predicted that in 1875 she would have 2 000 ounces of grain to sell and that she expected to profit the large sum of four thousand pounds for the year. She had achieved much using the outdoor system devised by Monsieur Alfred Roland in Switzerland.

Orders for trees flowed to Sarah in huge quantities and she received more requests than she could fill. This was in spite of the fact that she had a huge nursery of rooted trees and over 8 000 rooted cuttings of Milanese and Venetian, as well as an enormous quantity of seedlings.

With her success and the extraordinarily positive presentation being given the industry in the press, success seemed certain.

The half yearly meeting of the Victorian Ladies Sericultural Company was held on 4 June 1874 and reports were all very positive. The balance sheet was in the black and the first report on the education of grain imported from Switzerland was very satisfactory. A change occurred to the makeup of the Board on the resignation of Mrs Hackett. Her place was taken by Mrs Bull of Castlemaine. The directors were assured that the market in Italy was already assured. They were delighted to report on the favourable comments of the Italian Consul, Chevalier Marinucci regarding their product.

They reported that in New South Wales the government was offering "every assistance" to Charles Brady. He was to receive peppercorn rental of over 1000 acres of ground and would, with the expected sanction of the government, employ apprentice boys from public institutions to labour on his farm on very advantageous terms. No mention is made of payment to the boys, if any. They said Mr Brady already had 2000 ounces ready for export.

The Victorian Ladies Sericultural Company directors advised that they intended to send their grain to Europe as soon as the results of the spring education were a reality and ready to ship. They were particularly pleased with the result of the education at Mrs Neill's Mulberry Farm at Corowa, as distinct from company operations, and also at the Domain in South Yarra, as the quality of the grains and cocoons had "both been pronounced".

By that time eleven acres of mulberry trees had been planted on the Mulberry Farm site. An excellent description of the property appears in a letter written to *The Argus* by the manager of the Corowa branch of the Bank of Victoria, J.F. Daniell, in January 1874. He wrote in glowing terms of Sarah's work and the property.

> "A lady who, during many years residence
> of Corowa, has striven with all her energy to
> encourage every local industry, and to promote
> with admirable tact and cleverness, every
> scheme having for its object, either charitable
> benevolence or intellectual improvement.

> "She has planted Lucerne for mulch and many Ailanthus, Bamboo and Acacia for windbreak protection of the young trees. The grounds surrounding the dwelling house and the magnanerie are adorned with many foreign trees and beautiful flowers. Here the Elm, the Lilac and the Hawthorn recall to us the thoughts of many an English scene. Climbing roses trellised with vines bring France to remembrance while oranges and pomegranates in full luxuriance of bloom tell us we are far from home. The plantation of mulberry trees covers many acres and a large number of the mulberries are over 7 years old. The forcing frames contain at a rough estimate 20,000 seedlings and almost as many rooted cuttings."

Then follows a detailed description of the magnanerie:

> "It is one hundred and eight feet long and thirty feet broad, and very lofty – substantially erected from timber (probably cut from the local forest and raw un-sawn) and roofed with bark. The enclosure is a huge cage made of cheesecloth. The building, being open at the sides, permits a free current of air to circulate through it. The cost was about Two hundred and Fifty pounds. It will accommodate 400,000 to 500,000 worms."

Mr Daniell concluded:

> "There seems every probability that Mrs. Neill's indefatigable efforts will be crowned by complete success."

No examples of any of these plants remain on the site in 2016, nor is there much trace of the thousands of mulberry trees that she planted there. Several very old mulberry trees were found mid 2016 at the bottom of the farm property, near the river, by local residents Jan and Russell Black. If they are the original trees they are an astonishing 140 years old. This is possible as trees planted by John McArthur at Parramatta in 1796 were still alive in the early 20th century.

Although her plantation of the time consisted mainly of cape mulberry, it was Sarah's stated intention to concentrate in future on the *Bombyx mori* silkworm which preferred the white (*Morus alba*) mulberry and thus vast plantings of those would ensue.

Evidence appears in an item in *The Sydney Morning Herald* dated September 1884 that indicates that Sarah had named her property at Croppers Lagoon "Cardannah". This is repeated in an item in the *Deniliquin Pastoral Times* in an obituary of September 1884. There is no record of a property named Cardannah in the Corowa district. We speculate that this in fact was "Candaha", incorrectly spelt by a journalist. We believe it was named by Sarah in memory of her late husband John, who was closely involved in the British military force that invaded the city of Candaha (now Kandaha) in the first Anglo-Afghan war of 1839.

Chapter Eleven

*The Who's Who of Company Directors.
The Fine Arts. On the road with acting
and singing. Possums like mulberries!
Orbe on the mount in full swing.*

The Victorian Ladies Sericultural Company was initiated by Sarah and Jessie, especially inviting women to become shareholders. It was registered on 13 June 1873 under the presidency of Sir George Vernon, with secretary being W. Otter Esquire of the exclusive Yorick Club of Melbourne. All very proper, with the "right" people at the head; no doubt designed to encourage confidence in potential supporters of the venture. Sales of shares at four pounds each went very well.

The first directors were from prominent, wealthy families and were Charlotte Barker (wife of Dr William Barker), Frances Hackett, Caroline Lynch, Catherine Anderson, Caroline McGregor, Mary Wallworth, Florence Tripp, Jessie Grover and Sarah Neill. All affixed their signatures to the registration document. Now everything was in place for the expansion of the sericulture plan long ago hatched by Sarah. Briefly, as described earlier, it was the intention of the company to establish a number of branches throughout the colony, to recruit and train new participants from them as grainers and potential producers. The company would make the sales and distribute the money. A simple enough plan, perhaps comparable to the modern, but outlawed, pyramid sales scheme?

By the time Orbe Silkworm Farm at Mount Alexander was in the process of establishment, at Sarah Neill's Mulberry Farm at Corowa, the project was up and running. The magnanerie, completed in 1872, was in full swing by 1874, with silkworms in residence busily spinning their cocoons. The worms had hatched from the excellent, healthy grain provided by Charles Brady. Large quantities of leaves were being harvested from the plantations of mulberry trees. To maintain the health of the insects, absolute cleanliness, as instructed by Louis Pasteur, was a necessity and strict procedures were in place. Nearby was a large shed described as the leaf room and within a stone's throw was the residence. Scattered about were several other buildings necessary for a farm. It was a scene of much activity. The first year proved to be a disappointment and much of the product was lost when a shortage of food led to the deaths of thousands of worms. The local possum population had discovered an appetite for the leaves and wreaked havoc on the trees. The clouds of dust that descended on leaves, making them unsuitable for worm food, added to the problem. Leaf production fell way below what was needed and the season was largely lost. However, a small number of excellent cocoons had been produced.

As all this was happening, Sarah was busy with other community activities, mainly in support and participation of "The Arts".

In 1907, 94 year old Joseph Levin recalled her in his diary. Joseph was something of a legend in Corowa, coming from an old established family; his father had built the first flour mill in the town. The following extract is provided by the Corowa

Historical Society, holders of the copyright. Joseph wrote:

> "One of the old identities was Mrs. Sarah Bladen Neill of Mulberry Farm – and nothing in the nature of the history of Corowa would be complete without her name being mentioned. She was to be seen in the street with her turn-out almost daily. She made a great effort to start the Silk industry in Corowa and after travelling to the leading silk producing countries, she floated a company with five thousand pounds capital, grew mulberry trees and tried to induce people all over the district to do the same in order to get plenty of food for the silkworms, but very few people took it on. She kept a number of young women at her place teaching them all branches of Silkculture [sic] and gave lectures and had the girls with her demonstrating all branches of the industry, which were very interesting, but unfortunately for those concerned with it, the experiment was not a success."

Sarah was also instrumental in generously assisting with establishing the Corowa School of Fine Arts, a building which still exists behind the façade of the Memorial Hall in Sanger Street. The original building was a simple structure with whitewashed walls and a stage at one end, lit by kerosene lamps. The building was extensively altered in the 1860s, but parts of the original building are still visible. This includes the timber ceiling, which was rediscovered in about 1980. The hall, along with the Church of St John, stands as a type of monument to Sarah Florentia Bladen Neill.

One wonders how she found the time and energy for all of

her activities. She even led a touring theatre company which performed in a number of district theatres.

In his memoir recalling his days as a riverboat captain on the Murray, Augustus (Gus) Peirce wrote of this in his book "Knocking About", published in 1924 by Yale University Press. Gus incorrectly gives her the title of "Lady" and elevates her late husband to the British aristocracy as "Colonel Sir Bladen-Neill". (Many people mistakenly wrote his name as hyphenated, but as we know, it was one of his three given names.)

> "I made the acquaintance of Lady Bladen Neill. Her ladyship was then living in a bungalow at her large silkworm raising establishment at Mulberry Farm and she was spending much time, money and energy in forming amateur theatrical enterprises which she took on short tours, the proceeds of which were donated to various charities in which she was interested.
> "Having heard that I was something of an actor she invited me to the farm and easily persuaded me to join her company, which was composed of her friends and acquaintances, among whom were the Reverend Mr. Bing and Mr. Daniell the banker of the town."

Gus apparently had a soft complexion and was chosen to play the part of Princess Fatima in the comic opera, "Bluebeard". He wrote,

> "I was gorgeously dressed in one of her Ladyship's gowns ornamented by a beautiful diamond necklace. Her Ladyship was Sister Ann

and Mr. Daniell a ferocious Bluebeard. At the end of the performance I appeared in a Roman toga and gave Antony's address to Romana."

By pre-arrangement the performance was interrupted by the orchestra bursting into an Irish jig and the narrator danced a breakdown in a flying toga. The audience loved the comedy of it but it was "to the great disgust of Lady Neill who was very indignant of the most undignified performance".

Sarah took the troupe on a tour stretching over twelve days to Eldorado, Euroa, Benalla, Chiltern, Rutherglen, Wangaratta and Beechworth. The group travelled in a large horsedrawn coach. "Lady" Sarah travelled in a buggy driven by Gus Peirce. "We camped out and had a very jolly time," concluded Captain Peirce.

Sarah's activities in the theatre and general entertainment were many. In 1877 she organised a novelty charity entertainment in the Corowa Mechanics Institute, titled "Matters Silky". She took the opportunity to exhibit a gown made in England from silk from Mulberry Farm. It was a highly amusing evening and voted a great success with much singing and story telling, including an Italian song from Sarah herself, which, according to *The Ovens and Murray Advertiser*, "took the house by storm and she followed this with a most vociferous encore, the song 'Bother the Men' which evoked continuous applause. The event realised I hear, about Thirty Pounds". Sarah was a person of many talents.

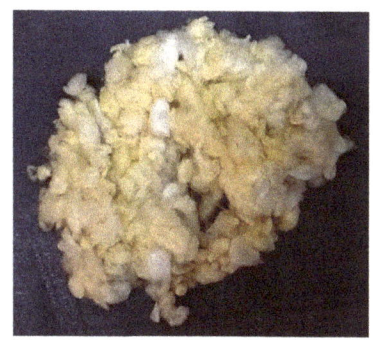

Above and below – the ruins of the cottage at Orbe sericulture farm on Mount Alexander, Harcourt, near Castlemaine

Raw silk from Orbe farm, Mount Alexander. Courtesy Castlemaine Art Gallery Museum.

A hank of spun silk from Orbe. Castlemaine Art Gallery Museum

Michael Cannon and Jill Smailes

Harry and Elizabeth Grover.

Statue of James George Smith Neill Wellington Square Gardens, Ayr, Scotland. Copyright wfmillar, used with permission.

Below is the plaque at the base of the statue.

The colours of the 40th Regiment at Foot.

The Vulcan which brought the Neills and the 40th Regiment to Melbourne in 1855.

Swindridgemuir, the Smith Neill home in Scotland.

The grave of John and Sarah in the Melbourne Cemetery.

The only remaining Mulberry tree on the site of the original Mulberry Farm at Corowa. Picture by [I]an and Russell Black.

The plaque of dedication recovered from the original St Johns Church, Corowa.

[C]roppers Lagoon at Corowa.

Saint Johns Church, Corowa.

[I]an Braybrook and Rev Rex Everett [o]utside Saint Johns Church, Corowa

Sue and Jane Broadmore at the Old Melbourne Cemetery.

A view of the Magnanerie at Corowa.

The unique roof structure of the magnanerie at Corowa.

Jan and Russell Black with Ian Braybrook at Corowa.

The shutter from Orbe, located at Faraday.

Chapter Twelve

Off to Europe again. Our cocoons and grain impress Italy. A lecture to London society. Sarah gains French staff. Harry and Jessie move to Orbe. Not enough leaves. More dust, frost and possums. The Mountain abandoned.

It was late 1875, and time for Sarah to make another visit to Europe and the United Kingdom to gather more knowledge. Sarah took with her samples of grain and the cocoons she had raised at Mulberry Farm. She went first to Italy making a number of calls on silk producers, leaving cocoon samples for examination, then on to England.

In London she appointed Miss Miles to establish a Silk Sales Depot for her product at number 7 Charles Street, Grosvenor Square, close to the place of her birth. Forever the publicist, Sarah wrote a letter to *The Times* advising readers that Miss Miles had arranged a demonstration of the education process using her grain. As a result, many visitors attended and were very impressed by what they witnessed. Thousands of silkworms emerged from the eggs, growing quickly to an extraordinary four inches in length. Experienced sericulturalists had never before seen such size and health in worms. The fame of the Australian product was spreading!

Later, April 1879, Sarah arranged for the re-titled "Australian Silk Depot" to relocate to even better premises at number 3 Charles Street Grosvenor Square; still in the heart of the best part of town, of course! Obviously catering for the well-off, the prices were high, even for those times. For example, knitting silk sold at one shilling and three pence per ounce, Stockings at ten shillings and sixpence a pair, scarves up to thirty shillings each and under-dresses "fitting closely to the figure" at thirty-two shillings.

A letter from the British Consul in Milan arrived in which he told her that examination of the cocoons their experts had received from her showed that sales of them would be easily achieved "to any extent in this important silk centre of Italy". That letter was followed by one from leading merchant Cavaliere Ferrarie of Verona. He told her that the quality of the cocoons delivered to him were so superior in quality that few people could recall seeing any to equal them for over twenty years. At this time another powerful advocate for Mrs Neill was also in London. It was none other than the Anglican Bishop of Goulburn, Right Reverend Stuart Thomas, who was well acquainted with her establishment at Corowa. He sang loudly her praises to all who would listen. Australia was being looked at as a country from which an inexhaustible supply of quality cocoons or grain might be obtained. The sky seemed the limit for her and Australian sericulture.

At that time Sarah was invited to give a lecture on "Australian Silk Growing" at the London Society of Arts. She was loudly applauded and the opportunity further advanced her fame and her profile.

She then widely circulated a document requesting assistance from anyone interested in the silk trade in England and the employment of women. The special objectives of the Victorian Ladies Sericultural Company were featured.

After visiting family she was off again to France to visit

some of the ailing sericulture farms. It was here that she was introduced to Monsieur Etienne Thibault, his wife and two other experienced men. It was another fortuitous meeting and had a significant effect on Sarah's work at Corowa. Among those she met was Thibault's brother-in-law Josef Gurein. All three men had been in the industry in France all of their lives and therefore had much hands-on experience. She promptly engaged them all to come and work for her at Corowa, with Etienne Thibault as foreman. The group was more than happy to leave the depressed industry in France and face a new challenge. It turned out to be a good move, as their knowledge was to prove invaluable. Everything was falling into place. What could possibly go wrong?

Meanwhile, the farm on Mount Alexander had been experiencing considerable difficulties.

In 1874 Jessie Grover, her husband Harry and four year old son Montague had moved to live at Orbe on the mount. They occupied most of the small cottage, with one bedroom going to the two young women who were there to learn the trade. The couple supervised the girls and several local labourers they hired. Their job was to plant and water the many thousands of mulberry trees and rooted cuttings brought from Mulberry Farm and the Domain, as well as plant thousands of seeds. By the end of August, twenty acres were turned into a fledgling mulberry plantation.

The trees were planted in rows ten feet (three metres) apart with a space of about six feet (two metres) between plants. The cuttings placed in frames for further raising and the seeds planted similarly. At the bottom of the gully, almost due north, the labourers excavated by hand a small dam. This was to be the sole water reservoir for the home and the plantation. The water was carried in buckets for several hundred metres up a very steep grade to the farm site. All was in readiness for that first education in December 1874.

The best of plans often go awry. The winter of 1874 brought heavy frosts that severely damaged the young mulberry trees. All of the leaves, the source of food for the thousands of worms expected to hatch, were lost.

Jessie and Harry assured the staff and the public in a statement to the press: "The plants are very vigorous and, given genial weather, they would soon bud out." Alas, they did not do so in sufficient numbers.

It was an urgent situation and, although the curators at the Botanical Gardens in Castlemaine and in Bendigo provided many leaves, and a few local friends helped out, there was not sufficient feed. As a result, the worms died and the first education was entirely lost. It was a severe blow to the infant enterprise and did not bode well for the future. In spite of the earnest efforts of Jessie and Harry Grover, assisted by the two young female students, the job was proving most difficult.

The next season was little better, with not only frosts to contend with. The kangaroo and possum population of the mountain had discovered that mulberry tree leaves made a delicious meal. The marsupials severely damaged the trees, killing many of them. Leaves were again far from sufficient and were again collected from the Botanical Gardens in Bendigo and Castlemaine, from a friend at Sutton Grange and from any available source. This was not enough; some thirty to forty pounds of leaves were required every day, a formidable amount.

It was hard work being a sericulturalist. In the season, work was constant. The day began at 5 am for the first feeding of the six for the day. The last was just before nightfall. In addition, the leaves had to be plucked from the mulberry trees, one at a time, being careful not to pluck too many. In that event the trees would die. The hardest job of all was cutting each leaf into small pieces for consumption by the worms. This was a most arduous and boring task requiring intense concentration. It was in fact exhausting.

As a result of the lack of food, many more worms died that season and only a very small harvest of cocoons was made. There was sufficient to hand-wind a small amount of silk and the quality of the resultant product was excellent. It was bad luck that the poor seasons had come at the outset of the enterprise.

The extreme difficulties faced on the mountain site were overpowering. Money was being sunk into the project and financial resources were dwindling.

To assist in this the board of the Victorian Ladies Sericultural Company voted to make a drastic change. Capital was to be increased from five to ten thousand pounds through shares at two pounds each, as opposed to the previous five thousand pounds capital with shares at four pounds each. The extra money was desperately needed.

It proved to be a losing battle and by the end of 1876 the company made the decision to shut down the farm and relocate all of the trees to Corowa where Sarah had arranged for land, probably part-owned by her, to be leased. It is recorded that the company leased seventy-four acres of land through a local agent, McKinnis and Company. Early records show that the Neill, Gray and Aitken partnership owned seventy-four acres just below Croppers Lagoon (Map of Corowa, drawn circa 1861, lot 135). This is probably the land referred to and presumably the company knew of Sarah's share in the ownership.

The buildings on the mount were abandoned. A newspaper

reported that, although the company was relocating to Corowa it had not yet decided to settle there permanently.

In February 1877 the *Mount Alexander Mail* informed the public of the Company's decision:

> "The Sericulture Farm at Mount Alexander has been broken up due to the unsuitability of the situation and the want of proper soil for the Mulberries. It has not however been all labour loss. Under the care of Mr. and Mrs. Grover, the quality of the silk produced there and taken to England, where it has been spun and manufactured into various fabrics, have been pronounced by competent judges as unsurpassable in quality. Judging by the samples displayed in Castlemaine on Saturday, one would be very apt to come to the same conclusion. These sufficiently proved the efficiency of the education the worms had received at the Farm. Leading drapers of the town pronounced the samples excellent. But as Mrs. Grover assured us the materials before us were made of 'rubbish', our wonder increased. This word proved to be a trade description for a particular kind of cocoon, indicative of the finest quality, rather an apparent contradiction in terms."

Be it rubbish, or good, bad or indifferent, the end had come for the bold sericulture experiment on Mount Alexander.

On reflection it is obvious to us that Mount Alexander was totally unsuitable for a plantation of mulberry trees. Even the trees indigenous to the area have difficulty surviving the harsh variations in climate – ranging from scorching forty-plus degrees to a freezing minus five or less. Not a single trace remains today of any of the 37 000 mulberry trees the Jessie and Harry Grover team planted there. It may well be expected that some would have sprung up from seed dropped from the fruit at the time. Alas, there are none. Any that may have grown from seed could not survive for long on Mount Alexander!

In a massive effort, a team of men, hired locally, dug up the remaining 7 000 trees still having life enough to suit transplanting and wrapped them in hessian, preparatory to having them transported to Corowa. The recently completed railway to Echuca made the job relatively easy. Loaded at the Harcourt train station they were unloaded at the bustling Echuca wharf. Here, the trees were loaded onto a riverboat for carrying to Wahgunyah. From the wharf they were moved by wagon to nearby Croppers Lagoon.

The ground had already been prepared and the trees were quickly planted in their new home. Although it was not the ideal time of year, the trees suffered little setback and recovered rapidly. Things were looking good at Corowa, but sadly, Orbe at Mount Alexander was left to decay.

It wasn't long before local farmers and others moved in to dismantle and carry away much of the buildings. Apart from a fairly substantial portion of one small granite building and the foundations of another, every piece of usable material was gone. The only current trace of the timber magnanerie is one framed roof vent shutter made of wood, installed on an old farm building at a nearby vineyard. The owner is keen to preserve it.

These days the ruins of the Mount Alexander silkworm farm are deep in the forest, under heritage protection, and the whereabouts completely hidden by bush. Although it is publicised on Parks Victoria literature and appears on social media, it remains largely unknown and rarely visited, except by a handful of locals. This is probably why it remains much as it was when abandoned and plundered by locals in the 1870s.

It is an eerie feeling to visit the ruins; visions of what it once was come readily – and with mixed feelings.

* *

Undoubtedly with regret and disappointment Jessie, Harry and Montague, now eight years old, returned to live comfortably in St Kilda. Montague, known to all as Monty, was sent to a private school. Jessie obtained a part-time position as social editor of the *Melbourne Bulletin* and correspondent to *The Queen* magazine in London, a prestigious position. She became probably the first female journalist in Australia. We believe she preceded the more famous Louisa Lawson by some months. Louisa was, of course, the mother of Henry Lawson, who is widely regarded as Australia's greatest writer. Jessie reported on social events such as the comings and goings at Government House on the Domain and the garden parties, charity events and bazaars hosted by society dames, and even the odd scandal or two.

Harry contributed occasional items to *Melbourne Punch* as a hobby. It seems however that he resumed his life of ease, principally engaged in monitoring his share portfolio and counting his money! He even purchased a racehorse, named it Jessie, entered it into the Melbourne Cup and lost. He was described in contemporary documents as a "Gentleman". To earn such a title, one had to be accepted under the British class system of the time as one worthy of the upper reaches of society.

Late in the 1880s the *Melbourne Bulletin* was close to broke and was absorbed into a rival publication. Jessie lost her job but continued to obtain writing work with several publications, using her pen-names such as Gladys, Iris, Hummingbird and Queen Bee.

Then the depression hit and in a flash Harry and Jessie lost all they had. They watched in horror as even their shares in Broken Hill Proprietary plummeted. Banks were no longer secure and many people were financially ruined. In a final blow for Jessie and Harry, the Red Lion hotel was condemned by the Council and shut down. They had retained ownership of the hotel, leasing it out. The rent provided an income stream and that too was lost. The Red Lion was eventually demolished.

Now the Grovers were flat broke and they were forced to leave their home in St Kilda and move into rented rooms. Jessie undertook charity work, obtaining food and clothing for the unemployed poor in her neighbourhood and opening her doors to provide many with food. She ensured that each child had at least one good meal a day. She was made a Life Governor of the Alfred Hospital along the way. Her work was largely unsung, but she did a wonderful job.

* *

Montague (Monty) Grover, the only child of Jessie and Harry, became famous in the journalistic world. In 1922, as editor, he founded a new style of newspaper in Melbourne, *The Sun News-Pictorial*, probably a world first. For the first time, news was accompanied by a large collection of photographs. This led to *The Sun News-Pictorial* being the largest circulating paper in Australia and Monty Grover becoming a journalistic legend. The annual Monty Grover award is still presented to a journalist judged the best by his or her peers.*

**

Things were certainly not going as well as Sarah would have liked, but her enthusiasm never faltered. In June 1876, according to *The Banner*, she was part of an "entertainment and lecture" at Corowa's Mechanic's Institute.

The reporter wrote:

> "It is scarcely necessary for me to say that the originator of the night's instruction and amusement was the indefatigable Mrs. Bladen Neill, who has already earned herself a world wide reputation in connection with sericulture. Mrs. Neill mounted the stage and gave a very interesting as well as instructive and amusing address on 'Matters Silk' and also exhibited specimens of the beautiful silk manufactured in England from cocoons grown by herself at Mulberry Farm, Corowa. The lecture occupied about half an hour in delivery and was listened to patiently by a very attentive audience."

Nothing, it seems, could dampen her energy, enthusiasm and belief in sericulture.

By 1877, there had been no great improvement in the fortunes at Mulberry Farm but Sarah had obviously made a change to the magnanerie. The *Corowa Free Press* of April 4 reported that – "the temperature in the magnanerie is now

---

* We are indebted to Michael Cannon's excellent book, "Hold Page One – Memoirs of Monty Grover", published by Loch Haven, for much of the above information

kept at forty degrees Fahrenheit, by artificial means, and the worms are thriving." This indicates that alterations had been made to the building to allow for better heating and cooling, along the lines of some magnaneries in Europe.

The report continued: "The season's operations at Mulberry Farm are fast approaching a most successful termination. The preparations for the education are well underway and the worms are almost at the cocooning stage."

Another important statement read, "We have every reason to believe that another sericultural establishment is to be started in the neighbourhood of Croppers Lagoon."

Indeed there was, the Ladies Sericultural Company had begun their move from Mount Alexander.

What was unknown, and therefore not reported, was that Sarah was about to relocate her entire enterprise to her land on the edge of Croppers Lagoon. Here she owned, in partnership with Aitken and Gray, several hundred acres which covered almost all the northern bank of the lagoon and beyond. The exception was a piece of about fifteen acres which was declared a public reserve. Sarah leased it for a time but soon cancelled it. The land remains a reserve to this day. In early times it was a popular picnic ground as described in a news item of the day when a St John's Sunday School picnic was held there.

The move from Mulberry Farm to Croppers Lagoon was a massive undertaking and involved the preparation of soil for the thousands of trees, many of them now mature and difficult to transplant. Although the past season had been dry, nevertheless the work of transplanting the trees proved very successful. Beginning in late winter, about 12 000 trees were moved, most of them very large and now six years old. Amazingly few of the trees failed to survive the transplant and continued with their normal spring growth. The ground had not been ploughed and properly worked, as time was a factor. Instead holes, two feet (60cm) deep and 3 feet (90cm) round, were dug to accommodate each tree. Rows were 20 feet (6 metres) apart. This was a huge undertaking with much of the work done by Chinese labourers. The trees were transported by several horse-drawn wagons.

The move also meant the erection of new buildings, including a cottage. The move was not finalised until the new rammed-earth magnanerie was constructed in late 1877. This was to be a different type of magnanerie to that previously used. It was described in an article in *The Leader*, 16 March 1878:

> "The magnanerie, built under the direction of French experts, is divided into two compartments, each measuring 24 feet by 20 feet. The walls are *piece*, or earthwork, formed by ramming tightly a mixture of clay and sand in a moist state into a long box and then ramming that mould and repeating the process until the desired height is obtained. The thickness of the wall is 2ft 6 in. and the roof is formed of thatch, laid on at a thickness of about two feet. The materials of the walls and roof, as well as their thickness, are chosen on account of their power to resist the severe heat of the summer and the choice is justified by the fact that the highest point registered on the thermometer throughout the present season was 78 degrees. The roof is so constructed that a chamber provided with numerous windows runs along the ridge of the building thus affording sufficient light without admitting the direct rays of the sun".

(The temperature range was established to be a remarkable 18 to 24 degrees Celsius.)

This description of the building matches almost exactly the building still standing there today, with the only exception being the thatch having been replaced by galvanized iron at some time.

This design was to give better air flow and a cooler environment in the heat of summer and allow for efficient heating by wood and coke stoves in the cold weather. It would maintain a more or less constant temperature, as had the revamped magnanerie at Mulberry Farm.

The newly available Carret stove was to be used; it was the most efficient stove available and, fired with coke and good dry wood, it would burn for up to nine hours. Precisely how this stove could heat such a building, with no ceiling and a roof half open to the world, we do not know. Perhaps the unexplained small openings in the lower walls, and a wooden pipe that appears to have been channelled under the floor may have had something to do with it.

It is interesting to note that Sarah and the Ladies Sericultural Company were the Australian distributors of the Carret stove and various newspaper articles included advice to readers where they could be obtained. Sarah never missed a marketing opportunity!

Because of his skill and experience, Etienne Thibault was the one responsible for the magnanerie design and he also oversaw the construction by Chinese labourers. There were many Chinese men in the district at the time due to their influx following the earlier gold discoveries. Sarah already employed a couple of Chinese labourers at Mulberry Farm and several of these men were given the job of building the new magnanerie. The quality of their work is illustrated by the fact that the building remained largely intact 140 years later, the principal defects being caused by the ravages of white ants and damage from a fallen tree, which toppled in a violent wind storm. This unique building is largely unaltered, except for the central door being closed off, and the roof thatch replacement. It is believed to be the last remaining building of its type in Australia.

Before long the move from Mulberry Farm was complete, but many of the mulberry trees were left behind, being too big to remove. It made good business sense for Sarah's private operation and that of the Victorian Ladies Sericultural Company to sit side by side, sharing labour and resources. At any rate Sarah would have been well aware that the move from Mount Alexander was the last roll of the dice for the company which was by now in serious difficulties.

Every day she was present to supervise work at the new site and the activities at the Ladies Sericultural Company site next door. At this time she received a bale of cocoons ordered from France by Etienne Thibault ready for the approaching new season. Mulberry Farm was abandoned.

As always, she was upbeat. She had managed to harvest some excellent cocoons at Mulberry Farm, which she had forwarded to the company's agent in London. The reaction from there had been positive and most encouraging.

The Duke of Manchester inspected the product and publicly spoke very highly of the silk reeled from the cocoons, suggesting it would be suited to use even by royalty! He would, he said, advise the Princess of Wales of the adequacy of the silk from Australia. On hearing this, a delighted Sarah ordered the remaining cocoons to be hand-wound and some excellent silk was the result.

Throughout 1877 newspapers continued to speak positively in support of Mrs Neill's efforts. Such expressions as "Progressing favourably" and "Thriving vigorously" filled many columns but, all along, things were not as good as they seemed.

The half yearly meeting of the company, held in Melbourne in December 1876, was not exactly positive. Money was in short supply and something had to be done. The directors recommended that the company be wound up and reformed, with increased capital of sixty thousand pounds, a huge lift from the previous ten thousand pounds. This would be raised by issuing 60 000 shares at one pound each.

The members present at the meeting agreed and the changes put in action by Jessie Grover.

Sarah, not long back from Europe, gave a stirring address to the meeting. She highlighted the opportunity for Australia to capitalize on the serious problems of the industry in France and Italy. She told of the offer from Milan to take twenty tons of cocoons at sixpence a pound. Tons, not hundredweights were required! She said:

> "The share price of one pound would not hurt anyone, even if the enterprise should fail, and if it succeeded the capital could then be increased, as I have explained, to one million pounds. In Western Australia something has been done by the government and the success of their effort is highly satisfactory."

She continued, at some length, to tell the audience that Mr Brocklehurst, the British MP and one of the largest silk manufacturers in Britain, had written a strong letter to the Australian government. In this he urged them of the advisability of supporting the movement. She told them of the fifty yards of beautiful silk cloth produced in England from cocoons from Western Australia, cocoons she had taken with her on her visit. She delighted the members when she explained the impact of the cocoons she displayed in Milan. "They were the best seen there for twenty years. The conditions are ripe for Australia to move." It was a glowing, confident recommendation for the company to push ahead and for members to dig deep in support.

She explained that her plan had now changed; instead of exporting cocoons and grain, Australia should send the completed product. This meant that reeling facilities would need to be established and people, mainly women and girls, taught the methods.

She told the meeting that proof of the future of the company was demonstrated in the one thousand pounds recently subscribed by her contacts in England and was available for use.

"I am going to teach some of the Chinamen on my farm at Corowa on the operation of reeling." She then gave her first public notice of her move from the site in Corowa. "A new site in Corowa had to be found as the present one (Mulberry Farm) was not suitable. And you ladies must not stop deputationising Ministers until they consent to accede to our wishes". No doubt meaning to pressure them to agree to financially support the industry.

She earnestly believed that the company could be made a great success, provided that "all worked together with energy to obtain the end".

**

Sarah Florentia Bladen Neill was held in such high regard that support for her generally remained strong. If political leader William Wentworth's proposal that Australia have an elite aristocracy system as in England had succeeded, Sarah would almost certainly have been made a Dame or another of very high rank.

There were a good number of sericulture companies and individuals experimenting with the production of silk in Australia in the 1860s to 1880s.

Indeed a report in the *Shoalhaven Advertiser* of September 1881 listed a good number of towns where sericulture was being undertaken. They were Melbourne, Ballarat, Ross Bridge, Rutherglen and Collingwood in Victoria. In NSW, Corowa, Albury and Tweed River; in Queensland, Ipswich; in South Australia, Adelaide and Gawler and in Western Australia, Perth, Caversham and Guildford. As well there were experiments happening in Auckland and Wanganui in New Zealand. Maldon also appeared on one published list of towns in Victoria but no evidence has been found of its existence.

One noted individual example was the determined effort by a woman in South Australia. Her name was Mrs Lindsay; we do not know a lot about her personally, even her given name.

She had been one of the women who studied under Sarah and Jessie for some months at Orbe at Mount Alexander and the Domain in Melbourne. Through this she had acquired a great deal of practical knowledge which she put into effect at the Parkside Lunatic Asylum, supported by the colonial government to the extent of supplying the accommodation. Little more is known and scant records of her work exist.

We presume that she had a position of some authority at the lunatic asylum. It was here that she constructed a simple magnanerie in a large room given her for the purpose. We also presume that she saw it as a suitable occupation for inmates at the asylum. She was also focused on drawing the attention of the public to the potential for sericulture in South Australia. Like others she emphasized the suitability of the industry for farmers, women and girls.

Mrs Lindsay had taken with her a quantity of the Roland regenerate grain given to her by Sarah. The worms resulting from this hatching were very poor and did not survive. On the other hand, grain provided to her by another woman from Melbourne, proved to be first class. It is likely that the supplier of the successful grain was Ann Timbrell of Collingwood.

Mrs Lindsay had eventually accumulated a good supply of quality grain which she acclimatised ready for a season of education. A burst of cold weather and a scarcity of mulberry leaves affected the resulting education and it was a failure.

Ultimately Mrs Lindsay was forced to abandon her work, principally due to a lack of money, plus insufficient mulberry leaves, public support and overall government disinterest.

Chapter Thirteen

A job for orphans? Meet the Premier. Help wanted. Financial struggle. You will be a millionaire!

1878 was to prove a watershed year for Sarah and the company.

The *South Australian Register* told of "Mrs. Neill's patronage of Sir Hercules and Lady Robinson, who are countenancing her attempts to establish an Orphanage Silk Farm at Corowa, on the Murray River, although we are not sanguine as to the support she will receive from the Parliament."

The reference to the Orphanage Silk Farm came from Sarah's suggestion that sericulture could be introduced into orphanages and other social welfare establishments to provide worthwhile occupation for the children. In fact she had pledged to operate such an amenity at Corowa if government support was forthcoming. Twice now Sarah had raised the matter of government support.

By now the Ladies Sericultural Company was struggling and financial ruin seemed inevitable. Sarah decided it was time for drastic action so she sought an interview with the Premier of New South Wales, James Farnell. It was granted, albeit with considerable delay and an apparent lack of enthusiasm. On 25 July 1878 she and Miss Osborne, the matron of Corowa

Hospital, met with the conservative premier James Farnell.

The meeting was, however, not taken very seriously by Farnell and his Ministers in attendance. *The Australian Town and Country Journal* coverage of the event tells the story well. It is slightly edited but gives a very clear indication of the attitude to women at the time – and to Sarah's proposal.

> "The wiser shook their heads and said they have never heard of such a thing before. Mr. Bowie Wilson was joined by the two ladies who had the temerity to venture into such a masculine region. The well known Mrs. Neil is good looking and dresses tastefully, has vigorous conversational powers and almost enough energy to set the Murray on fire. Her companion is one of those wise and gentle women who are the administering angels of humanity, whose beauty lies in the sweet expression that characterizes the faces of those who spend their lives in bearing others burdens. Mr. Farnell smiled benignly upon them, washing his hands with invisible soap in invisible water."

The report continued:

> "A letter previously written to Premier Farnell by Sarah was read out. The honorable gentleman tried his best to look interested and held a pen in his hand as if to take notes, but nothing came of it."

The meeting was a complete put down, of which the premier and his companions should have been thoroughly ashamed. But they were not, of course.

The letter referred to had asked the premier to establish a mulberry farm at Corowa to be worked by orphan children living in homes at the farm. It also asked that they obtain grain from the French and Indian governments. This project was to provide interesting, worthwhile work for the children and bring an income to the farm. Ultimately it would be self sufficient.

The premier did not conceal his disinterest. But Sarah persisted and opening a brown paper parcel she showed him specimens of cocoons, wound silk and silk yarn among other manufactured goods.

The *Town and Country Journal* continued its mocking article:

> "Mr. Farnell then caught sight of a pair of silken hose, which was almost too much for his nerves. He vehemently protested that he was no judge of such things. Mrs. Neill persisted in showing beautiful silken items like fancy scarves and ties. In addition to these products from her own farm, she showed yards of plain satin woven from silk grown by the West Australian Government."

The message there of course was: "The West Australian Government is supportive, why not you?"

Sarah went on to describe her farm at Corowa, gave a description of the soil and the testing of her cocoons in France, where a leading manufacturer declared, "Mrs. Neill you will be a millionaire and all your friends will be millionaires".

Mr Farnell then sought assurance that the disease that attacked the industry in Europe would not transpire here. Mrs. Neill said she relied on the virgin soil and the methods she used.

After considerably more forceful conversation from Sarah the meeting concluded. The premier assured her that he was favourably disposed to some of her proposals and he and his colleagues would consult on the subject.

Of course, nothing happened.

In July 1878, in Sydney, a further meeting was called for interested people to discuss sericulture. It was chaired by Sir Alfred Stephen who told of Sarah Neill's experiences. He described her as a person undertaking work which involved a large amount of fatigue and the expenditure of a considerable amount of money which had the welfare of others at heart.

He said,

> "The true philanthropist in new countries is the one who sets the thoughts and labour of men on profitable industries by which remunerative employment may be found for those otherwise be idle and poor. Mrs. Neill herself presents the matter in a much simpler and truer light. She states that the Ladies Sericultural Company has been formed 'for the purpose of developing the silk industry in Australia with the view to employing large numbers of women and children remuneratively'. Help was now sought from private persons and from governments because an industry of this type can scarcely be established without it. In every country where sericulture is flourishing the industry originated in a model establishment formed and supported by public money."

Such pleas continued to fall on deaf ears.

Chapter Fourteen

Vienna Exhibition. Dark clouds. When will the industry make a start? A win in Sydney. Almost a drought. Money bin empties. Sarah joins the Agricultural Society.

1878 began well for Sarah. A friend, the Commissioner for Queensland, Mr Clarence Hodgson, called to see her on his way to Europe. He took with him samples of silk produced at Croppers Lagoon and entered it in the Vienna Exhibition. It was extremely exciting for Sarah and her companions, as the silk won the prestigious first prize. In the same year, at a special ceremony, Sarah presented two white silken scarves made from the Corowa product, one to Bishop Thomas, the Bishop of Goulburn, and the other to his wife. There was considerable public acclaim and many congratulatory words for Sarah and positive expressions about the future of the sericulture industry in the colonies. As well, there had been the impressive reception of Corowa silk in England and further high praise from the Duke of Manchester, a good friend to Sarah. In a significant boost to potential investment, the renowned silk merchants, Lewis and Allenby of Regent Street, made it known that they had purchased one hundred shares in the Victorian Ladies Sericultural Company.

All good news, but dark clouds were gathering on the horizon.

Support for sericulture in the press was beginning to fade and such support was vital. *The Sydney Mail and New South Wales Advertiser* ran stories not in the best interest of Sarah Neill and sericulture. The premise was that it was a subject which promised success, which had been around for just too long with nothing to actually show for it. That was understandable. In particular, the lack of response by the Farnell government was a definite negative.

"When will the industry make a start and prove it is an industry suited to this colony?" *The Mail* continued:

> "For years it has, to use a sporting phrase, been at scratch. Long ago Mr. Brady (of Tumbulgum) told us about sericulture and Mrs. Bladen Neill enlightened us concerning the great future that was in store for Australian silk growers. It has been talked about in London, meetings have been held everywhere, companies have been formed, pieces of silk woven, land has been granted or leased at nominal rental and numbers of helping hands held forth; but still sericulture is at the scratch."

These were damning words from newspapers with wide circulation.

The demise of the venture at Mount Alexander did nothing to foster confidence in the future of sericulture either. The problem for Sarah and her supporters was that the issue was now public.

Questions were being asked and there was much lively community discussion. As a result, the newspapers contained more and more letters and information on the subject of sericulture, mostly negative.

One said emphatically of the Ladies Sericultural Company that it had made little headway and its best friend would admit that sericulture has not been successful. Another stated that the question of whether silk can be produced profitably had never been satisfactorily answered. Others ridiculed Sarah's idea of providing employment for orphans.

> "Mrs. Neil knows, and all colonists know, that orphan girls of this colony can readily obtain work without the aid of sericulture. Mrs. Neill's proposal may at once be set aside."

Another stated that experience had shown that such a business was precarious and the extensive growing of mulberry trees involved expenditure far above the means of the sericulture company. That was indeed the truth. However Sarah responded defiantly in a statement to the press that "the exportation of grain would pay magnificently".

Premier James Farnell entered the fray asking, if the industry was so profitable why more people had not entered into it. Sarah responded that they did not believe in new things, to which Farnell replied that it was hardly a new idea.

One newspaper stated that Mrs Neill had recommended Corowa as the best site for the government to grant her company land. She had, they said, conducted business there for years so it would be interesting to know what profits this land had turned out. The premier's support, instead of opposition, was vital to the future of the industry and his opposition contributed much to its eventual downfall.

In spite of the bad publicity Sarah remained steadfast in her determination to move the industry forward. She called a public meeting in July at 100 Pitt Street Sydney with a view to form a company in New South Wales. It was well attended by people well known in Sydney society and was chaired by Lady

Robinson. Among the others attending were Lady Stephen, Mrs St. John, Lady Hay, Lady Dease-Thompson, Lady Innes and a good number of other society leaders. Sir Alfred Stephen addressed the meeting, speaking in glowing terms of the industry and Mrs Neill. Lady Robinson undertook to write home to Baroness Burdett Coutts seeking support from her friends, and Lord Ronald Gower, who was present, offered to contact other aristocrats of his acquaintance in England.

The object was to seek support from the "right" people in the United Kingdom. Eventually, the meeting voted to form a new company, leaving the details to Sarah and, after some more speeches, the meeting terminated. But was it enough to stem the tide? The negative writing in the newspapers continued.

It seems that, as well as press editors and writers, a doubting public and a disinterested government, Mother Nature was conspiring against Sarah and her crew as well.

The summer of 1878 was hot, dry and dusty, almost drought conditions prevailed. This was precisely what Sarah did not need. As a result, the expected and hoped for successful and profitable breeding season, for her and for the ladies' farm next door, was a disaster.

As if the heat and dust were not enough, there was more. The possums were numerous and every eradication method the workers could devise did little to reduce their numbers. They bred like rabbits. Of all things, they had now developed a taste for the bark of the mulberry trees, not just the leaves. Hundreds of trees were either ringbarked or stripped of their leaves and as a result died.

Sarah and the ladies were devastated. There was little else to do but abandon the Victorian Ladies Sericultural Company, to cease operations. There was no money left, the trees were dying, the company was in debt and, probably more importantly, there was no support and little interest from the government. The plantations were a dismal sight.

Sarah paid off her staff and that of the company, retaining only Helen Stuart. It was a dreadful time for her but, in spite of her heartbreak, she did not surrender. Instead she continued to work on with her one assistant, determined to win. But it was an impossible task.

What she and the industry were crying out for was government help; serious recognition. If only she could get the government men to understand the golden opportunity that was at Australia's feet. Her firm belief was that the silk producers in Europe were in desperate need of healthy grain and cocoons. The evidence was that this was indeed an indisputable fact. Australia was in a perfect position to supply the needs of Europe. There was no doubt that if Australia could produce the goods, Europe would buy them – millions of pounds worth!

Unhappily, there was a strong feeling that sericulture was not to be taken seriously; it was something to play with, to watch the worms at work. It was also not considered to be "manly", more for women and children to muck about with. The offhanded way that Premier Farnell had treated the small delegation was proof of this.

Sarah continued to be a strong advocate for the industry and remained very active in her community. In 1879 she was admitted as a member of the Corowa Pastoral and Agricultural Society in which she had so often participated, winning many prizes for her silk. She made a number of constructive suggestions to the committee and expanded the number of categories and prizes offered by the society. Among her innovations were prize categories for such female activities as sewing, and culinary operations. These things, she said, should be recognised and would add to the attractions of the annual show.

Sarah's involvement in her community, from the time of her arrival until her premature departure, was considerable. She consistently promoted and participated in the various arts (her work in establishing the Corowa School of Arts is mentioned elsewhere) and she was instrumental in forming the Corowa Choral Society in 1878, the committee of which she was vice president. She arranged and participated in numerous live shows in the district, especially in Wahgunyah and Corowa but also in Benalla, Chiltern, Wangaratta and other district centres. As an example, she organized a huge fundraising concert for the enlargement of the Benalla Trinity Church. She once sang at a performance of a popular musical, "Norma", to raise funds for the two hundred and fifty pounds annual salary of the local vicar. She was always listed as a generous donor to local charities.

In 1880 she entered samples of her product in the Sydney Exhibition and again won first prize for silk. She repeated this at the Melbourne Exhibition in November. Her energy knew no bounds. Giving up was not in her nature.

The original School of Arts, Corowa

Chapter Fifteen

*Disaster strikes. Near death. A move to
Doctor Barker's in Melbourne.
A nephew is killed.
Lack of success.*

Early in September 1881, Sarah was entertaining two guests at her home at Cropper's Lagoon. One was sericulture enthusiast Everett Millais, son of a well known British artist, and the other, English Physician Julius Osborne, both of whom were visiting Australia. As they sat talking around the open fire, Sarah entertained them by playing selections on the organ. It was a pleasantly warm and friendly atmosphere.

Without warning a kerosene lamp fell from the top of the organ. It landed in Sarah's lap and instantly her clothing was aflame. The horrified men quickly fell upon her and managed to extinguish the flames, but not before Sarah suffered extensive, agonising third degree burns. At the tiny hospital Doctor Osborne did all he could, which was little compared to today. Skin grafting was in its infancy and the technique unlikely to be known in the colony at that time.

Rehabilitation was a long and agonizing process and, sadly, Sarah never properly recovered. The physical and mental scars left by such an experience are well known. Her energy and enthusiasm, not only for sericulture, but for life itself deteriorated. It was fortunate that a medical man, with knowledge of what best to do, was present when the disaster struck, otherwise Sarah would almost certainly have perished. On recovering, Sarah stayed on at her home, but she was never the same again; rarely seen in the town and becoming something of a recluse. No doubt she had been shattered by her near death experience. This would have been compounded by the lack of success of her brave venture into sericulture.

After a time, a shadow of the dynamic leader she had been, she moved from Corowa and took up permanent residence in her city home at 6 Barkly Terrace, East Melbourne, close to the doctor of her choice, William Barker.

**

On Saturday 31 July 1880 the following item had appeared in the *The Weekly Times*:

> "A fatal accident occurred ten miles from Corowa on Tuesday night. It appears James V. Neill, of Goombargama, left Corowa about dusk to return home, and he was not afterwards seen alive. The following morning Mr. Frederick Piggin, when on his way to Corowa, saw the horse of deceased feeding on the roadside. He took the horse to his own residence and later in the day the body of the deceased gentleman was found lying on the road, death having evidently been caused by coming into collision with an overhanging branch. Mr. Neill was the son of General Neill, of Indian celebrity, and the nephew of Mrs. Bladen Neill."

Frederick Piggin who found the horse was a well known auctioneer in the Corowa district. A magisterial inquiry, held soon after, confirmed the circumstances surrounding James' death, adding that it was a very dark night and that his forehead and temple had been crushed. It concluded that he "left a wife and child unprovided for". (We assume that "unprovided for" means he had made no will. No further details can be found and no record of a will is registered.)

The young man in question was Sarah's nephew, James John Vansittart Neill. James had married Alice Bristow and they had one child, Eric Vansittart who, as an adult, went on to a distinguished military career. How James, the youngest son of General James Neill, the British hero of the Afghan and Indian wars, came to be in Australia has not been properly ascertained. Most likely he was called upon to come to this country to assist with the administration of his uncle's will. Whatever the reason, he was living at Goombargama station near Corowa in 1880. This was quite possibly a property that the Neill, Gray and Aitken partnership owned and James

 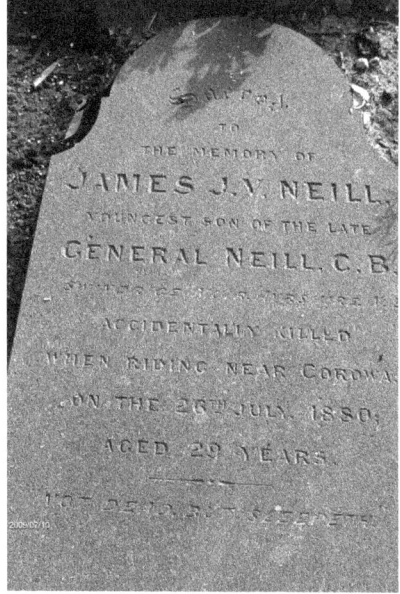

James Vansittart Neill. The headstone on his grave has fallen to the ground – it should be re-erected out of respect for the family.

was perhaps charged with managing it. That is a logical assumption, but it has not been ascertained.

James is buried in the Corowa Pioneers Cemetery, at the extreme back of the graveyard. His headstone has long ago broken midway and fallen to the ground. It is very simple, hardly befitting a man from the illustrious Neill family. It states only that he was James A. V. Neill, the youngest son of General Neill. The date of his death is shown as 29 July 1880, and his age as 29. There is no mention of his wife Alice or his son Eric. This appears to be rather odd but there are suggestions that he and Alice were estranged. The description of his father as "General" Neill is, of course, incorrect – he was a Colonel.

His death must surely have effected Sarah, especially given the circumstances of his death being so similar to that of her husband in 1859.

Eric Vansittart Ernest Neill with daughter Patricia, mother of Jane and Susan Broadmore. Colombo, 1919.

Chapter Sixteen

Death calls early. A lonely goodbye. The church benefits. Glowing tributes. The face for a five dollar note? Candaha is sold. The company is liquidated. Sarah's dream is ended.

What an unhappy life it had become for Sarah Florentia Bladen Neill. Her husband dead and she left with no child to comfort her, the dream of a huge industry for her adopted country, on which she had expended so much effort and sacrifice, lay shattered, her Corowa home and friends left behind, and her health ruined. She had been too ill following her burns to remain at Corowa for long, needing the best medical attention available to her. Her move to occupy her city home at 6 Barkly Terrace, East Melbourne would have been a huge wrench for her. No doubt she left the district and her home with a heavy heart, realizing it was almost certain to mean the approaching end of her life.

The brave Sarah breathed her last on 29 August 1884 aged just 56, her once dynamic energy finally expended. At this time she was being cared for in the nearby home of Dr Barker in LaTrobe Street, East Melbourne. Her death certificate describes her demise through mania and epilepsy. Sadly the names of her parents are listed as "unknown", her marital situation also "unknown".

She was buried with her husband, in the space reserved

for her in the grave at the Melbourne Cemetery, on the next Sunday afternoon. Only a handful of friends attended.

Many obituaries appeared in many newspapers, these edited extracts from *The Ovens and Murray Advertiser*, which was reprinted in *The Pastoral Times*, are typical. The author wrote first of her accidental burning and continued:

> "It may be taken for granted that the accident shortened her days. It became evident that a change of scene and air would be beneficial and she removed to Melbourne, placing herself in the care of Dr. Barker. A letter from Mr. J. Law, former manager of the Bank of Victoria at Wahgunyah, placed at our disposal says 'Her state of health for a long time past has indicated the approaching end. It came suddenly in the form of epilepsy and after the first attack daily increased in number until on Sunday she had seventy attacks. Her death took place at 5 p.m. last Friday – very quietly – she being mercifully spared consciousness for some time before."

James Law, then of Melbourne, was appointed the administrator of her will in April the following year. He was obviously a trusted friend. No details concerning the content of her will appear to be on record.

The obituary then paid tribute to her many contributions

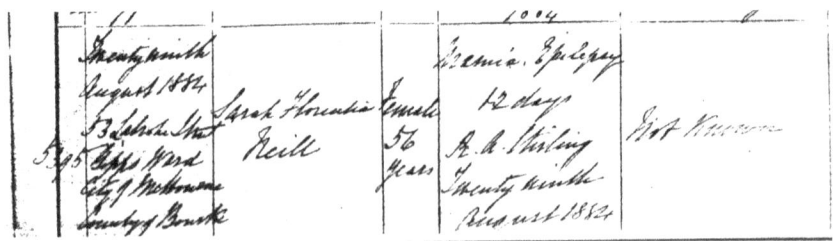

to Corowa, as well as her enormous efforts in sericulture.

> "She was a moving spirit in all things tending to the advancement of the township, a favourite in all classes. It would be strange if Mrs. Neill did not leave behind a large number of people who will speak kindly of her memory. Her death is generally regretted."

We could not put it better.

On 25 August 1885 a short item appeared in *The Ovens and Murray Advertiser*, Beechworth. It read in part:

> "Sale at Croppers Lagoon – On Thursday afternoon Messrs F. and A. Piggin offered for sale at Croppers Lagoon the property well known as the residence of the late Mrs. Bladen Neill. The land comprising in all about 350 acres. There was a large attendance and a total clearance effected."

Thus ended the dream, the life and the story of this wonderful woman, Sarah Florentia Bladen Neill.

* *

Sarah must have left a considerable estate, the distribution of which is unknown with one exception, that being a gift to her church. As stated, a record of her will is not available but presumably beneficiaries were family in the United Kingdom.

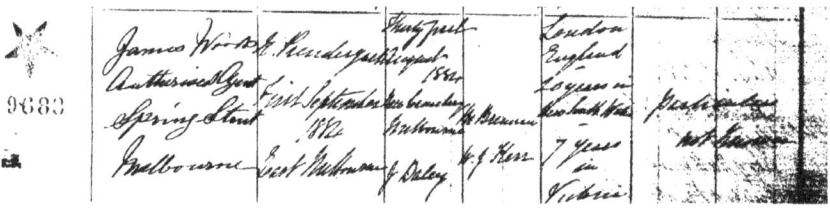

According to a notice in the *Sydney Morning Herald* in June 1885, probate was granted in the Supreme Court of New South Wales on April 18 that year to James Law, Bank Manager of Melbourne. Proctors were Klingender, Charsley and Dickson.

The sole item we know of, which would have been close to her heart was a bequest of four hundred and seventy-five pounds to her beloved St John the Evangelist Church in Corowa, the memorial to her husband, Colonel John Martin Bladen Neill.

There is an interesting twist to the fate of this bequest. The church committee wisely decided to invest this generous sum and asked the Bishop of Goulburn, Bishop Thomas, to invest the principal and hand over the annual interest yearly. Bishop Thomas arranged for an agent friend, a Mr Davidson to invest it in a mortgage on a small property in Goulburn.

Twice each year, in March and September, the resulting seven percent interest was received by the Corowa St Johns Church committee and put toward paying the salary of the vicar.

In 1895 Mr Davidson was declared bankrupt and apparently involved in some minor illegal activity. Through this action it was revealed that the mortgage had actually been paid in full seven years before in 1888 and the mortgage discharged. However the funds had not been reinvested. Mr Davidson had pocketed the principal but had dutifully continued to pay the interest in full when due. Consequently, nobody was any the wiser. Davidson's bankruptcy and his jailing meant that the payments abruptly stopped.

This was a serious blow to the church and, in particular, the vicar. His salary was reduced by the corresponding interest income of thirty-three pounds a year. At the urgent request of the vicar the new Bishop of Goulburn undertook to look at the matter and find out if the late Bishop Thomas could be held responsible or if anything could be done to retrieve the money. As expected, nothing could be done.

**

On 13 April 1893 the final chapter of the company closed. The Victorian Ladies Sericultural Company Limited was put into liquidation. The registrar general was advised by the company liquidator, E.H. Whiteman, that a meeting of members the previous day had the final accounts laid before them which showed how the company's properties had been disposed of.

The accounts showed assets of three hundred and fifty three pounds which were distributed as follows:

Writ issued against the company by Mr Grover – two hundred and two pounds.

Mortgage interest was ninety-one pounds and sales of property commissions – nine pounds.

Minor miscellaneous items, including legal costs, allowances to secretary etc. accounted for the balance. It was signed off by the liquidator. Why this writ was issued by Mr Grover is not explained but it may have been unpaid wages.

**

Many reasons have been put forward for the failure of the Victorian Ladies Sericultural Company.

In 1894, company expert and economist, Leonardi Pozzi, wrote an interesting and comprehensive piece on the reasons he believed caused the industry failure.

Of the Victorian Ladies Sericultural Company he said:

> "It was over-officered, subject to too many useless formalities and left behind a want of confidence

in one of the most important industries in Southern Europe".

He appears to place the blame for lack of confidence by investors with Sarah's company.

He continued:

> "In California it was a success. Why not follow them? The big hold up or disadvantage is 5 years before the trees give return. It does not explain the failure."

Chapter Seventeen

Finding a Magnanerie

Finding and eventually identifying the Magnanerie at Corowa proved to be difficult and troublesome at times. The following account of the process will probably be of interest.

The extracts from *The Mount Alexander Mail* collected and collated by Harcourt historian, the late Hedley James, were the inspiration behind our eventual "discovery" of the rare magnanerie at Corowa. From Hedley's work we knew that Mrs Neill and the Victorian Ladies Sericultural Company had set up a silkworm farm in Corowa, but we knew no more. So in August 1990, we made the 300 kilometre journey from our home at Harcourt to Corowa, to see what we could find out. We were encouraged by our fellow committee members of the Harcourt Valley Heritage and Tourist Centre. It was really the beginning of a fascinating journey into the past that was to take us over twenty-five years to complete!

At Corowa we didn't have any special objective; we simply wanted to get any information we could – anything at all to add to our meagre knowledge. Our first port of call was the Federation Museum where we met the late Val Swasbrik. Her knowledge of local history was extensive but that of sericulture in the area was limited. Val advised us to go first to the church of St John and then to the library.

At the church we met Canon Bill Ginns. He was helpful with information about the church but he was not able to throw any light on the subject of sericulture. We then proceeded to the library where they held copies of the local newspaper. Again, everyone there was helpful and, to our surprise, we were allowed to examine the musty, but carefully stored copies of *The Corowa Free Press*. These went back to the era we were focused on – 1860 to 1890.

After several hours searching we had unearthed some really valuable new information. Most importantly we found several references to Mulberry Farm and a mention of sericulture at Croppers Lagoon on the Mulwala road.

We were also able to examine and photocopy a compact booklet produced many years before. Among other things of interest, it contained a small piece about sericulture and, lo and behold, a photograph of Sarah Neill dressed in an elegant gown which it said she wore when presented to the Queen of Portugal. It said that the gown was made from Sarah's own Corowa silk! That was pretty exciting!

But our knowledge was still very scant.

Val had given us the phone number of Mrs Dixie Leslie, formerly of *The Corowa Free Press*, so we spoke to her, hoping she may have learnt something through her job. She was happy to talk to us but knew nothing on the subject of Mrs Neill. She confirmed that there had once "supposed to have been" a sericulture place on the Albury road about two kilometres from town. We went to take a look at the approximate site but there was nothing there to indicate it was ever Mulberry Farm. We took a few photographs of the site which contained a couple of large sheds.

While having dinner at the RSL club that evening we met with local couple, Russell and Jan Black who were seated at the next table. As a result we were invited to their home for coffee the next morning.

Before we left town they drove us to the site of Mulberry Farm and assured us that this was definitely the place and it was always known as Mulberry Farm. The land concerned had once been owned by a family member.

Our homeward journey took us past Croppers Lagoon where we now believed a sericulture farm once was. As we drove past Riverside Horse Stud, Marilyn called for me to stop. In the distance she had seen a peculiarly shaped building. "That's it. That has to be a magnanerie," she cried excitedly.

Indeed, behind a group of trees was a strange looking building. It deserved investigation, so we drove onto the property to have a look. Greeting us in the house yard was Len Rhodes, the owner of the farm. He welcomed us warmly and was happy for us to take a look at the building.

Standing there before us, surrounded by what we took to be white mulberry trees, was a most unusual building. It's unique roof structure, rammed earth walls and general appearance suggested that it was what we had been looking for. We believed that we were on the site of a sericulture establishment so what else could it be? It was a style of structure we had never seen before and, although we didn't even have so much as a picture of one to compare it with, we were confident that we had found the magnanerie. In reality, all we had was Hedley James' description taken from his local newspaper. We told Len Rhodes that although we had no direct evidence, we firmly believed that the building was a magnanerie and of considerable heritage importance.

He said, "I've often wondered what it was and a few times I was going to pull it down." We begged him not to even think about it and he happily agreed. We left without any doubt that we had discovered the Corowa magnanerie.

At home I wrote a lengthy article outlining what we knew about Sarah and sericulture in Corowa and sent it to *The Corowa Free Press*. They published it but it didn't draw the hoped for response. We had hoped that some reader would come forward with information, especially about the building, but nothing happened.

* *

Several months later, passing through Corowa on our way home from Sydney via Albury, we called on Len Rhodes again. Although we still had no proper description of a magnanerie, we were convinced that we had found one. We advised him that we believed the building should be put on the NSW Heritage or Historic Buildings Register. He agreed it was a good idea and said he would look into it. We were satisfied that Len would never allow the building to be pulled down. He was quite interested and enthusiastic about its history and possible heritage value.

At that time our lives changed and we put thoughts of Corowa, Sarah Neill and sericulture aside; other things took priority. On occasions we would drive through Corowa and pass the old building in the distance, seeing it still standing, hoping that it was being preserved by Len. Maybe it had even been listed on the Heritage database!

The interest was always there but fifteen years passed before we were able to resume our delving into sericulture and Sarah Neill.

In mid 2016, now in retirement and on a government pension, we decided to re-visit the subject and returned to Corowa to take up where we had left off so long ago. As we drove past the Rhodes property we were delighted to see in the distance that the old building was still standing, albeit with a bit of a lean.

The next morning Ian drove out to the Croppers Lagoon farm to check if it was okay to come back and take some photos and have another look. It was then that we met Len's daughter Dyonne. She was in the process of buying the farm from her father, so it was fortunate that he went to visit at that time. She showed considerable interest in what Ian had to say; in fact she had been doing a bit of research on the old building herself. We were invited to come out again later in the day, which we hastened to do.

Our hope was that she and Len would agree to us trying to have the building placed on the Federation Shire Heritage List and more importantly on the New South Wales list. Not everyone is willing to allow this sort of potential for intrusion on their privacy. Happily, Dyonne agreed.

Our next visit was to the Church of St John where the vicar, Rex Everett, was pleased to show us through the building. It is quite beautiful, with numerous magnificent stained glass windows. He showed us the plaque erected in honour of John Martin Bladen Neill and gave us an interesting booklet on the history of the church.

We next visited the Federation Shire offices and spoke to Adrian Butler in the Planning Department. He was supportive of our intention and referred us to Peter Kabaila, the Shire Heritage Adviser, who was newly appointed and looking for local projects. Peter was part-time and only visited Corowa once a month. We emailed him some details and photos we had taken and arranged to meet with him at a future date.

We went home to Castlemaine and by email made arrangements for a meeting with Peter and Dyonne on site in July.

We were pleased when Peter proved to be quite enthusiastic and resolved to put his weight behind the move to have the old building listed. After a careful inspection he recommended certain remedial work to be done and Dyonne agreed to start the ball rolling. This work included removal of some trees, the roots of which were causing damage to the walls, and repairing or replacing guttering. He also suggested the erection of a small veranda around the perimeter to stop further water erosion on the mud walls. Inside, urgent work was needed to prevent the collapse of a section of roof badly damaged by white ants.

Peter then told us he would write a report to the Federation Shire and begin the process for New South Wales Heritage Listing. We were delighted and offered to help by providing the supportive information we had accumulated.

Things were looking good but Marilyn and I soon became quite troubled. When we started to collate the material for this book we realised with a shock that we had no real evidence that the building was a magnanerie. It became obvious that the detailed description of the building on Mount Alexander that we had read in *The Mount Alexander Mail* in no way corresponded with what we had at Corowa. The Mount Alexander building was twice the length, with open sides, surrounded by a veranda; there was no resemblance whatsoever. Then we uncovered another description, in *The Argus* of December 1873, of the magnanerie at Sarah's Mulberry Farm. Again it bore no resemblance to the Croppers Lagoon building.

By this time we had serious doubts. Maybe the Corowa building was in fact some other type of specialist building?

Getting quite desperate, we looked for possible helpers. Who could positively identify the building? Should we look overseas perhaps? They had lots of experience in Europe?

We traced two people we had read of who had considerable knowledge of sericulture, Ueli Ramseier, CEO of the Swiss Silk Producers Association and Professor Panomir Tzenov, Director of the Department of Agricultural Science in Bulgaria, an expert in sericulture. We sent a series of photographs of our hoped-for magnanerie and expressed our concern that the building was not what we had thought it to be. Both of them were non-committal, saying only that "it could be a magnanerie". This did nothing to boost our confidence.

What sort of fools would we be if we had given false information to the owner, Dyonne Rhodes, and Peter Kabaila the heritage adviser? We reluctantly emailed them suggesting that they may have inadvertently been led up the garden path by us, but we emphasized that we still held hope that we were correct. It was a very worrying time; our credibility was at stake.

This remained the situation for about two weeks. During this time we scoured every conceivable source for a description of a magnanerie that matched what we had. There was none.

Very early one morning, unable to sleep for worry, Ian was on the internet when he discovered a forgotten item in *The Leader* of 16 March 1878, referred to previously. It was written by their "Travelling Journalist" who had only recently visited the newly established plantation and farm at Croppers Lagoon. He said that Mrs Neill had adopted the "warm system" i.e. maintaining a temperature range between 65°F and 75°F. For this she needed a different type of rearing room (magnanerie).

The description he gave of the newly built structure matched precisely the Croppers Lagoon building we were looking at! The only difference was that the roof was then thatched, "2 feet thick", with rushes harvested from around the lagoon, not corrugated iron.

This information provided us with tremendous relief to say the least. We were then able to advise Dyonne Rhodes and Peter Kabaila, that it was full steam ahead. This description was later confirmed in a 1914 letter to *The Weekly Times*, discovered much later, by Eric Vansittart Neill of South Yarra, son of James. Coincidentally Eric lived on Punt Road, the street where his grandfather John had died in 1859.

They were not mulberry trees around the magnanerie, by the way. We were told that they were white figs! No mulberry trees were anywhere to be seen.

The End

Appendix

When we first saw the sericulture farm ruins on Mount Alexander they were completely surrounded by a pine tree plantation. This was planted about 1953, intended as a future fund provider for the Castlemaine Technical and High Schools. We believed that the trees posed a threat to the ruins; a tree may fall on the remaining building and cause serious damage.

We approached the responsible authority of the time, the Department of Conservation and Environment in Bendigo,

The ruins on Mount Alexander before the pines were planted

asking that the trees be removed. The initial response was disappointing. In their opinion the trees required another ten years of growth to maximize the financial return to the school and should remain.

We then met with the Technical School principal, Alan Clifford, and on hearing our story he readily agreed that he would recommend that the pines should be removed. On notification of the school's decision the department moved in and all the trees were cut down and carried away. The threat to the ruins was gone.

These days the remains sit in a clearing on a grassy hillside, completely obscured from view from the road. Very few people know of their existence and, if they do seek it, locating the place is almost impossible without a map. Such a map is published by Parks Victoria but is not readily available. This has ensured that the ruins remain untouched and almost exactly as we saw them in the 1980s.

Some years ago Marilyn joined a small group in scattering the ashes of Harry Grover amid the ruins on the site at Mount Alexander, the place where his father and grandparents resided and worked so long ago. It was a moving event, attended by Harry's wife Betty and a handful of friends.

* *

Whatever became of the buildings left at the original Mulberry Farm on Albury Road Corowa is a mystery. No trace is to be found today. Probably the usable material was removed for use at Croppers Lagoon and the remainder spirited away for recycling at other properties or otherwise crumbled away. There remain the few remnant mulberry trees at the rear of the property, found by Russell and Jan Black, twisted and gnarled. Of the elms, lilacs, hawthorns, bamboo, trellised vines and roses planted by Sarah, there is nothing.

As for the magnanerie we found at Croppers Lagoon, it is expected that it will be safely preserved; in the hands of a caring owner and listed for posterity on the New South Wales Heritage Register.

In 2016 we met with Jane Broadmore and her sister Susan at the Melbourne Cemetery. They are granddaughters of Eric Vansittart Neill, and the great-granddaughters of James (who died at Corowa) and Alice Neill. Eric had three children, Patricia, James and John. Patricia married James Broadmore and they are the parents of Jane and Susan.

We visited the grave of Colonel John and Sarah Neill and had an enjoyable and informative discussion over several hours. It was a pleasure to meet actual descendants of the original Neill family in Australia.

* *

Sarah Florentia Bladen Neill's scheme was unique. It failed because the people, and especially our governments, did not understand it; nor could they grasp the huge, profitable opportunity and the enormous potential for Australia that her plan offered at that time. If only the governments and the people had acted when Sarah, and others, first proposed a sericulture industry, we may well have been the world's foremost producer of silk.

In 1992, when the Australian government considered Sarah's image for a place on the five dollar note, it could not have gone to a more deserving person.

Afterword

Some brief history of sericulture in Australia.

Mulberry trees were first introduced into Australia soon after colonization, both for the shade they gave and the delicious berries they produced.

In 1796 John McArthur, regarded as the founder of the sheep industry in this country, imported and planted a number of trees at Parramatta. Some of these were still alive in the early part of the 20th century, demonstrating that they have a long lifespan.

In 1824 the Australian Agricultural Company was established in England for the purpose of setting up sericulture in New South Wales. Although some trees were planted, the venture went no further.

In 1840, T.S. Mort, President of the NSW Agricultural Society, began an experimental farm at Eastwood, near Sydney, planting 4500 mulberry trees. However, insufficient capital caused the venture to fail.

In 1842 a leading surgeon, Dr R. Jameson, advocated sericulture as a means of providing employment for young people of either sex. He was generally ignored.

Six years later, a Mr Beuzeville set up another silkworm farm at Eastwood, but it failed due to poor quality trees and, ultimately, lack of finance.

In the 1860s and up until late in the century, Charles Brady pushed very hard for the establishment of the industry. Born in Kent, England in 1819, he ran into financial strife and was bankrupted in 1852. He moved to Australia and lived at Curl Curl, near Manly in Sydney. Here he developed an interest in silkworm farming and in the 1870s imported silkworms from Monsieur Roland in Switzerland to commence experimental breeding of a disease-resistant strain of worm. Throughout the 1860s and 1870s he was heavily involved, published pamphlets and books and delivered lectures to any group that would listen, including the New South Wales Agricultural Society. He exhibited the results of his endeavours at every opportunity.

In 1869 he was again declared financially broken, with liabilities exceeding one thousand pounds and assets of two hundred pounds. His estate was sold to cover some of his debts.

Eventually, through his persistence, the government leased him 1200 acres of land on the Tweed River for him to continue his work in sericulture. The rent was set at one shilling per year. He planted thousands of mulberry trees and built a magnanerie and other necessary buildings. But finally he was forced to give up, blaming, probably correctly, the lack of interest and financial support from the government. No trace of his enterprise appears to remain. Charles Brady was regarded as the foremost authority on sericulture in Australia.

Affleck and Howard of Albury spent a huge amount of money and effort in the 1870s to establish their silkworm farm but, in spite of early success, disease killed every one of their worms. The idea was abandoned.

In the 1870s Signor Chevalier Bruno applied to the Minister for Lands for a long term lease of five thousand acres of land near Benalla for the purpose of sericulture. He claimed that he wanted to bring a colony of Italian workers, experienced in sericulture, into the country to establish a large industry. They would raise the worms and complete the entire process, even to the manufacturing of silk. What followed was considerable controversy.

To accommodate his request the minister had the crown land concerned reclassified as grazing land and Bruno was issued with a grazing lease at sixpence an acre. What a bargain that was! In fact it drew attention and a group of politicians opposed the lease, stating sericulture use was not within the meaning of the Act, which allowed for grazing only. Skullduggery was suspected and the minister was placed under quite a cloud.

There was then a lot of going to and fro, with Bruno claiming that he had already spent 2500 pounds on the project. An inspection showed that he had clearly not spent even an eighth of that sum. He also claimed that he had ordered 40000 trees from a supplier in North Collingwood (probably Ann Timbrell) and was obliged to pay her. The upshot was a Royal Commission being called for. Whether it went ahead is not recorded, but apparently it did not.

Whatever the outcome Signor Chevalier Bruno was the loser and the project never got off the ground.

A document held by the State Library of NSW shows that, in 1891, the Colonial Secretary, Henry Parkes, granted a group of Italian residents of New Italy in NSW a large area of land, and even paid a manager to supervise the men and carry out the work. After a promising start, it too failed through lack of knowledge.

In 1893 a NSW government report into sericulture stated that the industry was ideally suited to the colony and should be established, but only with government backing. The

Government Gazette announced that under the Crown Lands Act, the Governor, Sir Robert Duff, declared sericulture to be a purpose. Whatever that meant! Of course, considering long-lasting government disinterest, the report was shelved and ignored.

As late as 1918, Sir David Hennessy, formally asked the Minister for Lands, Mr J. Reed, on behalf of his wife Lady Hennessy, President of the Sericulture Society, to grant land to Mr H.A. Khat, a Syrian businessman. Mr Khat was attempting to establish large scale sericulture in Victoria and needed government assistance. He proposed to set up interested people, particularly men returning from the war, on a thousand acres, divided into twenty-five acre blocks. They would primarily grow mulberry trees but also walnuts, olives and almonds. Improvement to each small block would require five hundred pounds, of which Mr Khat was prepared to contribute substantially. Sir David said that the idea deserved favourable consideration.

Mr Reed asked that a submission from Mr Khat be made in writing and "as silk culture was a matter of interest to many women; no doubt we have a lot to learn." Not surprisingly, no record of any action being taken by Mr Reed is evident.

It is a fact there were numerous attempts at establishing a sericulture industry in every Australian state. For a variety of reasons none of them was successful. The principal reason appears to have been lack of finance and only a government was in a position to provide that.

**

Interest in sericulture continues today. A small establishment is operated by Sarita Kulkarni not far from Melbourne, in South Gippsland. Her work was recognised by the Federal

Government Research and Development Corporation in the early 2000s. Following that, she had some success with breeding; her trees are doing well but she has experienced difficulty in importing breeding stock. She also had her funding cut and it appears that government disinterest has once again thwarted any development of sericulture in Australia. In the nineteenth century all efforts failed, principally because of a lack of government support, essential to start a new industry.

Nevertheless, sericulture still has promise of being a worthwhile and profitable enterprise, as demand for silk is still high.

In the popular Margaret River region of Western Australia a successful silkworm farm tourist attraction is in operation. Visitors can see silkworms in action and examine the silk produced on the site.

There is definite potential for such a business in the Corowa district, capitalising on its historic connection with the industry.

Late in 2016, a small group of young Castlemaine mothers formed an organisation named Little Habitat Heroes. Assisted by Connecting Country, a group dedicated to re-vegetating much of the Shire of Mount Alexander, their purpose is to re-grow the area surrounding the ruins with indigenous plants. This is under the direction of the government authorities concerned. It is a long term project and takes into account the stringent conditions imposed under heritage legislation. Hopefully this project will ensure the proper preservation of these valuable ruins, which grow in value with every passing year. To date (2017) they appear to have been neglected, and recommendations made in 2001 have not been attended to.

Silkworms are still a popular novelty in primary schools and feature in numerous children's books.

What is Sericulture?

We won't get too technical, keeping it simple. Sericulture is in fact part agriculture and part horticulture. It is both farming and fruit growing; therefore it is rather exclusive!

Silk production sounds like the sort of thing many of us did as kids at primary school; mucking about with silkworms and watching what they did. It never amounted to much for most of us and we never did see any silk as a result. It was just a bit of fascinating fun!

The caterpillars, or silkworms, feed only on mulberry leaves – sounds like a pretty boring diet, but they love it and digest the leaves in huge quantities. They eat their way non-stop through several stages until they eventually spin cocoons of a single thread in which to hide until they emerge as moths. Most of us are aware of the process, but to gather the silk the creature has to be killed before it emerges as a moth. This was customarily done by applying heat through submersion in boiling water, rather a cruel method you'll agree. Sometimes heat is applied by placing the cocoons in the sun in the heat of day, but the hot water treatment gets far better results.

The spun silken thread is quite amazing – a single fibre so fine, so delicate and yet so strong, and it extends for an unbroken half kilometre or more.

After killing the worm the thread that makes the cocoon

can then be unwound by machine (when it was once done by hand) and then used to manufacture the most glorious and expensive garments.

Silk production, or sericulture, still has a vital place in the economy of many countries and is a very serious and profitable industry. It remains the highest priced natural fibre with demand ever increasing. At the same time, production is falling. Australia imports about fifty million dollars worth of silk products each year, a figure which is rapidly growing. We are well placed to produce our own silk but competition from established producers, such as China, France, Switzerland and Italy, would make it difficult to be viable. We could probably produce enough for our domestic needs, but the risk makes it unworthy of investment it seems.

China is a traditional producer of silk and currently provides about three-quarters of the world's requirements.

Some recent experimental work has been done in Australia but without any real success or enthusiastic take-up by governments. Some small scale efforts continue but, without serious government support it is unlikely, now as then, to develop its full potential.

The Australian silk produced by Sarah Bladen Neill, Ann Timbrell, Charles Brady and a handful of others in the 1870s and 1880s was actually world class. Its failure to gain government and business acceptance at the time was undoubtedly a serious error.[*]

* *

---

[*] We acknowledge the information lifted from a paper, The Silk Road by Adrienne Ferreira 2011, *What is a Silkworm?*

Following is a brief description of the life and death of a silkworm.

A female silkworm moth mates with a male and lays several hundred eggs. She dies almost immediately and her mate very soon after. At the right temperature of about 80 °F (25 °C) the pinhead sized egg hatches into the silkworm. This takes about seven days and the worm then eats constantly; it has an insatiable appetite consuming large quantities of mulberry leaves. It grows quickly and by day eleven or thereabouts it sheds its first skin, which is replaced with new. In another four days it sheds and replaces that skin and in a further six days does so again.

At about thirty-five days it is mature and can be up to four inches (100 millimetres) in length. It ceases to eat and then starts to discharge its silky fibre. After three days it encloses itself completely in a cocoon of the fibre and is now at the chrysalis stage. The silky fibre is ready to harvest.

At this stage the would-be moth cannot be allowed to emerge, as this would destroy the cocoon. It must be killed, usually as previously described, but sometimes in an oven.

Some are allowed to live to continue the life cycle. In that case the moths emerge and immediately the males and females mate. The female lays her eggs and dies and the male dies soon after so the cycle of silkworm life continues.

Index of People and Place Names

40th Regiment 22, 25, 27, 28, 32, 34

A

Abraham, King. *See* Munangabum
Acclimatisation Society 40, 51
Adelaide 79, 113
Affleck, Thomas 81, 147
Age, The 24, 34
Agricultural Society
   Corowa 123
   NSW 146, 147
Aitken, J.H. 35, 59, 101, 107, 127
Albury 81, 113, 138, 147
Alexander the Great xv
Alfred Hospital 105
Anderson, Catherine 92
Anglo-Afghan war 91
Apennine Mountains 53
Apsley Place 33
The Argus 26, 40, 77, 78, 86, 89, 140
Auckland 113
Australia Felix xv
Australian Agricultural Company 146
Ayr, Scotland 29–31

B

Bank of Victoria 89, 130
The Banner 106
Barker
   Charlotte 92
   Dr Edward xvi, 59
   Dr William xv, xvi, 92, 129, 130
   Madeline xvi, 59
Barkly, Sir Henry 28, 34
Barkly Terrace 37, 80, 126, 129
Barnweill 29, 47
Barry, Justice Sir Redmond 33
Beaumont, Mr 72
Beechworth Sericulture Company 50, 72
Benalla 49, 50, 96, 124, 147
Beuzeville, Mr 146
Bing, Mr 95
Black, Dr 46
Black, Jan and Riussell xiv, 91, 136, 144
Blight's Quarry xvi
Botanical Gardens
   Bendigo 100
   Castlemaine 100
   Melbourne 40, 41
   South Australia 51
Bowen, Sir George 76, 78, 86
Brady
   Charles 41, 45, 51, 60, 71–72, 78, 89, 93, 120, 146–147, 152
   Sir Antonio 47, 57
Bristow, Alice 50, 127–128, 145
British East India Company 30
Brocklehurst, James 48, 111
Brocklesby station 36, 43
Broken Hill Proprietary 105
Bruno, Chevalier 147–148
Bulgaria xiv, 47, 141
Bull, Mrs 88

Burke and Wills Monument xvi
Burns, Robert 29
Butler, Adrian 139
Byng, John (Field Marshall) xv

C

Cairo 47
Calvert, Lady Lucy 48
Candaha 91
Cannon, Michael 22, 106
Cardannah. *See* Candaha
Carret stove 109
Casey
  Mr 79
  Theresa 59
Castlemaine. *See* Mount Alexander
  Botanical Gardens 100
  Diggings xv
  Little Habitat Heroes 150
  Technical and High Schools 143
Ceylon (Sri Lanka) 30
Chapman, Tom 43
Chiltern 86, 96, 124
Chinese (Chinamen) 108, 109, 112
Clifford, Alan 144
Cole, Jack 84
Collingwood 113. *See also* Timbrell, Anne
Coppin, George 39, 59
Corowa xiv, 23, 84, 113, 121, 150. *See also* St Johns Church; *See also* Mulberry Farm; *See also* Croppers Lagoon; *See also* Agricultural Society, Corowa; *See also* Neill, James Vansittart
  Choral Society 124
  Finding a Magnanerie 135–144
  History 42–43
  Home on the River 35–39
  Honour Avenue 36
  Mechanics Institute 96, 106
  Orphanage Silk Farm 115, 117
  Pioneers Cemetery 128
  School of Fine Arts 94, 124
  Vines 82
Corowa Free Press 106, 136, 138
Coutts, Baroness Burdett 122
Croppers Lagoon 43, 59, 144. *See also* Vienna Exhibition
  Candaha 91
  Charles Cropper 43
  Neill, Gray and Aitken 101, 107
  Sale 131
  The Magnanerie 70, 136–140
  Vines 82
Curl Curl, NSW 41, 51, 147

D

Dalry, Scotland 31
Daniell, J. F. 89, 95, 96
Davidson, Mr. 132
Dease-Thompson, Lady 122
Deniliquin Pastoral Times 91
DePass, John 59
Dja Dja Wurrung xvii
Docker
  Elizabeth 59
  Rev Joseph 49
Domain, The 41, 69, 88, 89, 99, 113
  Magnanerie 76–78
Dublin 32
Duff, Sir Robert 149
Duffy Act xv

E

East Melbourne
  Barkly Terrace 37, 80, 126, 129
  Grey Street 62
Echuca xv, 103
Eldorado 96
Ellery, Mr 72, 75
Emu Inn 22

Eureka 28
Euroa 96
Everett, Rev Rex 139

F

Farnell, James (NSW Premier) 115–117, 121, 123
Fawkner, John Pascoe 33
Fellows, Judge Augustus 87
Ferguson, Sir James 46, 51
Ferrarie, Cavaliere 98
Ferreira, Adrienne 152
Fitzroy. *See* Old Colonists Homes
Forest Creek xv
Francis, Mr 79

G

Gaul, Dr Alexander 26
Gawler 113
Genoa 54
Ginns, Canon Bill 136
Goggin, James 24, 63
Goombargama 126, 127
Government House, Melbourne 76, 77, 104
Gower, Lord Ronald 122
Grasse 54
Gray, George 35, 59, 101, 107, 127
Grosvenor Square 97
Grover
   Harry 22, 23, 144
   Jessie 21–24, 58, 62, 69, 72–74, 92, 99–100, 102–105, 111
   Montague (Monty) 69, 99, 104–106
Guildford 113
Gurein, Joseph 99

H

Hackett, Frances 59, 88, 92
Hamilton Spectator 86
Handberg, Allan xiv

Harcourt xv–xvi, 59, 69, 72, 103, 135
Hart's List 32
Hay, Lady 122
Hennessy, Sir David 149
Heritage Register
   NSW 138, 140, 145
   Victoria xvi, 40
Hobsons Bay 32
Hodgson, Clarence 119
Hotham, Governor Charles 34
Howard, John 81, 147
Hume
   Andrew 43
   Elizabeth 43
   Hamilton 43

I

Innes, Lady 122
Italy xiv, 42, 51, 53, 87, 89, 97, 98, 111, 152

J

James, Hedley 135, 137
Jameson, Dr R. 146
Japanese grain 46, 51, 53

K

Kabaila, Peter 139, 141, 142
Kandaha 91
Kellow, Mr 75
Kerferd, Alfred 86
Kew xvi, 25, 35, 36
Khat, H.A. 149
Kingstown. *See* Dublin
Kooyong 25
Kulkarni, Sarita 149

L

Law, Mr James 130
Lawson, Louisa 104

The Leader 108, 141
Leanganook xvii
Leslie, Mrs Dixie 136
Levin, Joseph 93
Lewis and Allenby, London 48, 119
Liarga Balug xvii
Lindsay, Mrs 113–114
London 31, 57, 97
   Exhibition 42, 51
   Society of Arts 98
Lucknow 29–30
Lusitania 47
Lynch, Caroline 92
Lyons (France) 54, 57
Lysander 32

M

Madras 30
Maldon 113
Manchester, Duke of 47, 110, 119
Manly, NSW 147
Margaret River, WA 150
Marinucci, Chevalier 88, 89
Marseilles 54
Mayfair, London 31
McArthur
   Edward 33
   John 91, 146
McGregor
   Caroline 92
   Doctor 25, 26
McGuire
   Elizabeth 22
   Pat 22
   William 22
McGuire's Crossing 22
McKinnis and Company 101
Melbourne 113
   Cemetery xvi, 130, 145
   Domain 113
   Exhibition 72, 124
   Zoo 41

Melbourne Bulletin 104, 105
The Melbourne Club 59
Melbourne Cup 104
Melbourne Punch 104
Metcalfe Council 65, 66, 73, 76
Middleton, W.C. 42
Milan 54–57, 98, 111, 112
Millais, Henry 125
Mills, Lady 48
Mitchell, Major Thomas xiv–xv
Montague, George 48
Mount Alexander xiii–xvii, 140, 143.
   See also Leanganook
   Granite xvi
   Orbe 93, 99, 113
   Sericulture 21–24, 62–71, 76–77,
     81, 102–103, 107, 110, 120
   Shire 150
Mount Alexander Mail 21, 64, 69,
   102, 135, 140
Mulberry Farm 50–51, 53, 59, 62,
   70, 79, 84, 87–91, 93–96, 97, 99,
   106–107, 109–112, 136–137, 137,
   140, 144
Munangabum xvii
Murray River Railway xvi

N

Neill 43, 127. *See also* Smith Neill
   Eric Vansittart 127, 145
   James George 29, 30
   James Vansittart 30, 126, 127
   John Martin Bladen 21, 25–34, 48,
     132, 139, 142
   William 29, 30
Nerevy, J.F. 38
Neville, Lady 48
New Italy, NSW 148
New Zealand 80, 113
Nightingale Fund 33
North Eastern Railway 62, 84, 86
Novi 54

O

Old Colonists Homes 39
Orbe
  Mount Alexander 75, 76, 93, 99, 103, 113
  Switzerland 55–57
Orbe, Switzerland 75
Osborne
  Dr Julius 125
  Miss 115
O'Shea, Patrick 25–26
Otter, W 92

P

Parkes, Henry 148
Parkside Asylum 113
Parramatta 91, 146
Pasteur
  Louis 54–55, 93
  System 54
Pastoral Times 130
Peirce, Augustus (Gus) 95–96
Perth 113
Picnic Gully 69
Piggin
  F. and A. 131
  Frederick 126
P&O Line 57
Portugal 54
  Queen of 136
Pozzi, Leonardi 133
Princess of Wales 47, 110

Q

The Queen magazine 104
Queensland xiv, 113
Queen Victoria 47, 48

R

Ralph
  Edward 32
  Everel 32
  Sarah Florentia 31
Ramseier, Ueli 141
Red Lion Hotel 22, 105
Reed, Mr J 149
Rhodes
  Dyonne xiv, 141–142
  Len xiv, 137–139
Riverside Horse Stud 137
Roberts, Tom 36
Robinson, Sir Hercules and Lady 115, 122
Rockhampton Bulletin 78
Roland, Alfred 52, 55–57, 59–60, 63, 75, 80, 88, 147
Rome 53
Rutherglen 96, 113

S

Scotland xiv, 29, 57
Shepparton 23
Smith Neill 29, 47
  Caroline 30
  William 29–30
  William Francis 30
South Australia xiv, 113–114
South Australian Register 87, 115
Spain 54
Sri Lanka. *See* Ceylon
Standish, Frederick 59
Stephen, Sir Alfred and Lady 118, 122
St George Church 31
St James Church, Melbourne 23
St John Ambulance Service 39
St John Church, Corowa 43, 48, 94, 132, 135, 139
St. John, Mrs 122
St Paul's Church, Middlesex 31

Strangford, Viscountess 47
Stuart, Helen 52, 62, 79, 123
Sun News-Pictorial 105
Sutton Grange xiv, 74, 100
Swasbrik, Val 135
Swindridge 29, 47
Switzerland xiv, 88, 147, 152. *See also* Orbe
Sydney xiv, 80, 118, 121, 146, 147. *See also* Brady, Charles
Sydney, Exhibition 124
Sydney Mail 120
The Sydney Mail and New South Wales Advertiser 120
The Sydney Morning Herald 91, 132

T

Tasmania xiv, 80
Thibault, Etienne 99, 109, 110
Thomas, Bishop Stuart 98, 119, 132–133
Ticonderoga 32
Timbrell, Anne 39–42, 51, 59, 114, 148, 152
    Silk Farm 51
The Times 97
Town and Country Journal 76, 116–117
Tripp, Florence 75, 92
Tumbulgum 40–41, 120. *See also* Brady, Charles
Tweed River 113. *See also* Brady, Charles
Tzenov, Panomir 141

V

Valiant, Colonel TJ 32
Venice 84, 87
Vernon, Sir George 92
Vienna Exhibition 42, 119
Volunteer Corps 21, 28

Von Mueller, Baron 40–41
Vulcan, HMS 32

W

Wahgunyah 43–44, 86–87, 103, 124, 130
Wallworth, Mary 92
Wangaratta 49–50, 82, 96, 124
The Weekly Times 126, 142
Wentworth, William 113
Western Australia xiv, 41, 79, 111, 112–113, 150
White, Captain Hans 25–26
Whiteman, E.H. 133
Witt, Mr (Minister for Lands) 86
Wodonga 62, 86

Y

Yorick Club 92
Yorkshire 40

www.ingramcontent.com/pod-product-compliance
Lightning Source LLC
Chambersburg PA
CBHW051548010526
44118CB00022B/2617